Tropical Trees: Propagation and Planting Manuals

Volume 3

GROWING GOOD TROPICAL TREES FOR PLANTING

Written by K A Longman

Illustrated by R H F Wilson

January 1998

Commonwealth Science Council

Tropical trees are increasingly seen as a valuable renewable natural resource. They maintain and improve soil fertility, and provide protection from sun, wind and heavy raindrops. Trees also yield a great range of important products, play a crucial role in many farming systems, and form the base of the food-chain for numerous animals. Their presence is clearly essential for the survival of people in the tropics.

Despite these many vital roles, tropical trees continue to disappear about ten times faster than they are replaced, which threatens the life-support systems of many human communities. Yet the forests and savannas, farmland and woodlands where they thrive could be managed sustainably, providing soil protection, supplies of products and other benefits in perpetuity.

Now tree planting projects are springing up all over the tropics, and manuals are needed because:

- some agriculturalists thought that farmland should be completely cleared
- foresters concentrated on a few introduced species, grown in pure plantations
- economics was equated with maximising short-term gains
- knowledge of tree biology did not underpin land-use
- experience in one locality did not reach other regions
- without trees, the land cannot support increasing human populations

and also because:

- in the past, many kinds of trees just came up by themselves
- information is still lacking about growing most local species
- most of them are still 'undomesticated'
- seeds may not always be available
- just one of a number of potential problems could lead to failure

Tropical Trees: Propagation and Planting Manuals provide practical, illustrated guide-lines, based on general scientific understanding and local experience. The main aim is to encourage the growing, planting and care of trees on any site, by anyone, at any scale. The series covers all stages from genetic selection through setting up a tree nursery to planting and successful establishment in the field (see inside back cover). Also included are sources of further information and examples of check-lists, record sheets and worked examples of calculations.

The Manuals are spiral bound so that pages can easily be photocopied for use in the field. We hope that the series will stimulate translations into other languages and the writing of sheets specifically for local use.

PREFACE

This is the third volume in the series: *Tropical Trees: Propagation and Planting Manuals.* It deals with a vital stage in the increasingly urgent work of replanting trees in the tropics - growing good planting stock. The first two volumes are concerned with propagation by rooting cuttings and by germinating seedlings, while this one is about the planning and running of a successful tree nursery. Whether one is growing a few plants in a temporary site, or many thousands in a permanent nursery, the growing conditions affect the numbers, quality and size of the young trees, and when they will be ready to go out. One of the keys to whether they will survive and grow well after planting is the type of root system produced in the container or nursery bed. Others are proper nursery handling, protection and care, particularly at critical stages.

Trees have existed within and recolonised tropical farmland for many centuries, and so it might seem that there would be no need for manuals. On the contrary, precisely because many species were once self-replenishing, there is usually little or no experience of propagating them. At the same time, there are large areas of bare, degraded land in the tropics, without enough parent trees to provide seed for natural regeneration or for planting. Another reason is that running a tree nursery is one of those apparently straightforward undertakings in which any one of a range of problems and difficulties could arise. Unsatisfactory planting stock is frequent, and poor growth and failure to survive can occur because a single point has been missed in a well-run nursery.

Manuals are also needed because less experimental research has been done on tropical trees, compared for instance with some of the main cereal crops, and much of it is scattered and unavailable to the grower. As a result, there is a lack of underpinning of nursery practice with theoretical understanding of tree growth, particularly in physiology, microbiology and genetics. The results of practical trials may also be unknown outside the country where they were carried out. Yet a great diversity of what are effectively still 'wild' tree species need to be grown for planting on a wide range of different sites by a variety of people. In these circumstances, explanations of relevant information and experience are clearly necessary, put in the form of practical procedures and advice.

In this Manual, planning, setting up and running a tree nursery are covered in detail, with emphasis on the points to look out for in each procedure. As a basis for this, the general principles of tree growth are outlined in non-technical language. What cells are and how roots, stems, leaves and branches grow are explained, together with the maintaining of the tree's water balance, the absorption of nutrients, and the production and movement of substances. As in the other manuals, there are several sheets on experimental research, including the setting up, assessment and analysis of experiments with young trees in pots.

Only 25 years ago, it was thought that tropical trees had few mycorrhizal associations; nowadays it is recognised that these are very widespread, playing vital roles in the nutrition and probably the water relations of most trees. The effective recycling of nutrients in agroforestry and other woodland ecosystems depends greatly on mycorrhizas and nitrogen-fixing nodules, and it is likely that inoculation in the nursery would save on costly fertilisers and benefit field establishment in many more tree species than the few so far studied.

Genetics has progressed only as far as provenance testing in some of the commonly planted tree species, but is clearly of much wider importance than this. Because the need for more trees is so urgent, it would be a great pity (though understandable) if it was assumed that plants of any origin will do. In fact, since most trees are 'undomesticated', the potential for large genetic gains is far greater than for crop varieties that are already highly-bred. Indeed, some of these can lack the broad tolerance of extreme conditions built into older, traditional mixtures of agricultural strains. In both cases, genetic diversity clearly needs to be conserved, but for trees this should go hand-in-hand with domestication, which can be hastened by alternative propagation techniques. Then superior sets of genes will be retained and passed on, both in current planting and each time regeneration occurs in the future.

Sadly, many of those who recognise how important trees are, and would like to replant them within local landscapes where they are badly needed, lack both money and technical advice. Yet simple hints, cheap methods and additional species are now available, so that most problems can be foreseen or overcome. Readers are encouraged to look for ways of spreading the information in this manual to such potential growers, including the photocopying of relevant sheets, and their translation into other languages. It is also hoped that the manuals will stimulate the writing of sheets on individual tree species or particular regions.

Acknowledgements are made for the use of photographs and diagrams as a basis for some of the drawings in this Manual, including several from Fuller (1983), Goor and Barney (1968), Mung'ala *et al.* (1988), Napier and Robbins (1989), and Shanks and Carter (1994) - see sheet C 61-A, D. I should like to thank many colleagues, in the tropics and in Europe, for their help and stimulation. Suggestions from readers for additions and improvements, and ideas for Volumes 2 and 5, would be welcome.

Alan Longman
24 Orchardhead Road
Edinburgh EH16 6HN
U.K.

February 1997

The author, Dr Alan Longman, has worked in the West African tropics for 9 years, teaching plant physiology, doing research on trees and planning and developing a large tree nursery for vegetative propagation. He has taught and supervised students at all levels, and his research into the growth and development of trees included studies of flowering and phase-change, rooting of cuttings and early genetic selection, and the effects of day-length and temperature on shoot growth and bud dormancy.

Starting as a forestry worker in 1949, Dr Longman has been a Senior Lecturer in the University of Ghana, Principal Tree Physiologist of the U.K. Forestry Commission and a Charles Bullard Fellow at Harvard University. He began a research project on tropical trees at the Institute of Terrestrial Ecology near Edinburgh, Scotland, which has now been running for almost 25 years, with linked projects in several tropical countries.

He is on the Editorial Board of *Biologia Plantarum,* reviews scientific papers and books, and is an Honorary Fellow of the University of Edinburgh. In retirement, Dr Longman is writing the five Manuals in this series, and he also wrote the scripts on which the video accompanying Manual 1 was based (see inside back cover). Plans for the future include four more videos to accompany Manuals 2 to 5.

Contents
Growing Good Tropical Trees for Planting

Collecting wildings and soil.

INTRODUCTION

Introduction C 1
- *why grow tropical trees?*

Why are tropical trees important?

Because they play many crucial roles, especially in:

(A) creating and maintaining conditions in which other living organisms can thrive (D10-16 in Manual 4); ***and***

(B) providing numerous everyday products (D 33-40) that people need.

In spite of this, trees are disappearing very rapidly throughout the tropics, which is undermining the support systems for human life. Large numbers of young trees are needed as replacements, in order to restore land that has become degraded (D 22).

But don't young trees just come up by themselves?

Sometimes yes; but more often no.

Why not?

Tropical trees may fail to *regenerate naturally* (D 1-2 in Manual 4) for many reasons, including:

(a) lack of enough parent trees to provide pollen and seeds (Manual 2);

(b) site conditions having become harsher (Manual 4);

(c) repeated clearing, cultivation and/or burning by people;

(d) excessive browsing by animals; ***and***

(e) strong competition from aggressive weeds (Manual 5).

As a result, planting is now generally needed to supplement the natural regeneration of trees.

How do tropical trees actually benefit people?

(A) ***Indirectly,*** *by protection from sun, wind and heavy raindrops:*

(1) greatly reducing the blowing and washing away of fine soil particles (Manual 4);
(2) discouraging wholesale erosion of the soil on slopes;
(3) protecting lower land, rivers and lakes from sudden flooding and excessive silt deposits;
(4) ameliorating local, regional and probably global climates;
(5) shading the soil, so that it does not get so hot and dry, but is a favourable environment for the roots of young trees to establish and grow (C 11);
(6) encouraging the close associations between tree roots and micro-organisms (C 30-32) that lead to effective recycling of nutrients;
(7) providing better conditions for many farm crops and animals (D 3, D 21 in Manual 4); ***and so***
(8) maintaining the potential for ***sustainable management.***

(B) ***Directly,*** *by producing many items that may be:*

(1) essential for the lives of local inhabitants;
(2) sold in local and city markets, supporting people who collect, transport and trade in them; ***or***
(3) exported, which **should** benefit all those involved.

What kinds of products?

Hundreds of different kinds of items (D 33-40 in Manual 4), including for example:

(a) foods and medicines for humans and forage for farm animals;
(b) materials for building houses and fuel for cooking;
(c) wood for diverse agricultural and domestic uses and for export.

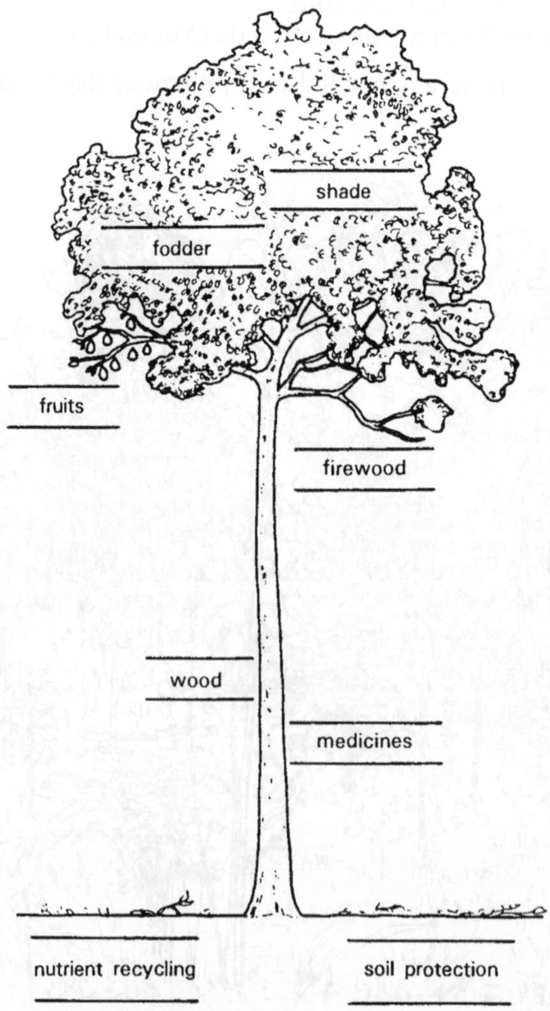

Could we really go on having these things for ever?

Yes! They were freely available from natural woodlands for many centuries.
Bare, degraded sites, now covering more than 20% of tropical land, can no longer meet human needs, but properly managed trees could do so in perpetuity.

Then why are trees disappearing so fast?

That's a question many people are asking. There are many answers, including:

(1) taking into account only a few of the multiple functions of trees;
(2) misunderstandings about:
  (a) how trees grow (C 10-15); ***and***
  (b) how tropical vegetation functions (D 10-16 in Manual 4);
(3) people concentrating on:
  (a) exploiting for timber; ***and***
  (b) clearing the land for agriculture and industry; ***instead of***
  (c) sustainable management of a valuable *renewable* resource;
(4) shortage of money, by individuals, businesses and governments; ***and***
(5) the fashion during recent decades for basing decisions on short-term economic indicators (D 1 in Manual 4), rather than the benefits of longer-term investment.

Aren't a lot of trees being planted?

Not enough, so far. Tropical trees are disappearing about **ten times faster** than they are being replaced.

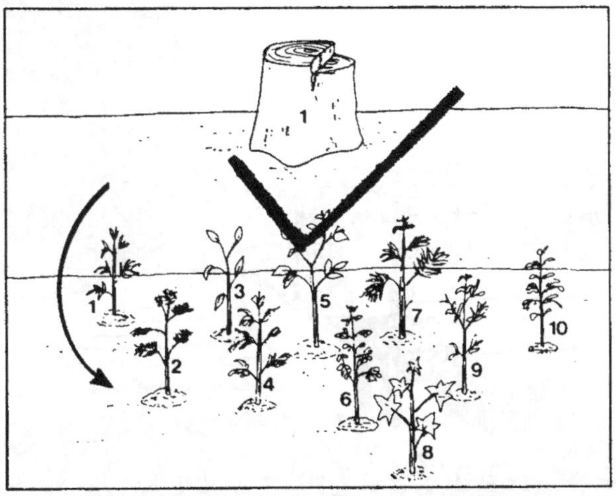

What is needed is at least **ten times more trees planted than cut,** to achieve a net rebuilding of the resource, allowing for those young trees that will not survive.

But surely it's no problem to plant more!

In one way, no, since growing and planting young trees is not hard to do; ***but***
In other ways, yes, because successful establishment of trees depends on solving a series of problems (C 3). Here are some examples:

(a) **Land tenure** - frequent problems about who owns the land, and whose trees they would be;

(b) **Expectations** - the various people involved may have conflicting ideas about how the land should be used (D 5 in Manual 4), and might not be familiar with the idea of planting trees;

(c) **Planting stock** - enough *good* plants (C 4) are not always available;

(d) **Thought before planting** - if the young trees are to thrive on a particular site, choices need to be made about the kinds of tree species and their usefulness, and the most appropriate planting patterns and methods to use (Manual 4);

(e) **Protection** - the young trees need looking after if they are not to be choked by weeds, eaten by animals, attacked by pests and diseases or killed by fire (C 40-48 and Manual 5).

What else do I need to consider when growing trees?

(A) Appropriate genetic origins of the seeds or cuttings (C 5 and Manuals 1-2);
(B) The general principles that govern the growth of trees (C 10-15);
(C) Where to put the tree nursery (C 2, C 20-26), and how to run it (C 50-54);
(D) Giving the young plants a good start (C 4, C 30-34);
(E) Getting them to the planting site in good condition (C 47 and Manual 5); and
(F) Planting them well (Manual 5).

Can you give a few general hints?

(1) Draw on local experience of growing woody plants, and look out for demonstrations about tree nurseries, useful publications (C 61-62), or other sources of help (C 53).
(2) Don't assume it will cost too much - small nurseries near the house may not involve much expenditure (C 2). For larger nurseries, there might be a financial incentive scheme, usually dependent on the successful establishment of the trees in the field.
(3) Try new things out on a small scale first, and don't be put off if you don't succeed at once.
(4) Remember that although most steps in growing trees are straightforward, each of them needs attention if the risks of losses and damage are to be minimised (C 4, C 60).

Introduction — C 2
- do I need a tree nursery?

What are tree nurseries?

Places for raising good young trees, which are then planted out where they are to grow.

Is that the only way of growing trees?

No - trees can also become established by natural regeneration (C 1; and D 2 in Manual 4). For *artificial regeneration,* nurseries are generally used, although for a few species you could also consider the possibility of:

(1) collecting wildings (Manual 2) and planting them directly in place;
(2) direct planting of leafless cuttings where they are to grow (A 4 in Manual 1); ***or***
(3) direct sowing of seeds (Manual 2).

Wouldn't such direct planting be easier?

Occasionally yes, if one of these methods has been well tried out for the particular tree species (as with various hedging and shade plants grown by method 2); ***but***
Generally no, for most kinds of trees, because:

(a) they are not suited to methods (1) or (2);
(b) method (3) typically requires a lot of seeds, to allow for failures; ***and***
(c) protecting directly planted young trees may be difficult (Manual 5).

Tree nursery near the house.

So what does the tree nursery really do?

Looking after the young trees first in a nursery can allow planting stock to be raised that is much more likely to establish well than if directly planted.

But with direct methods you wouldn't have to disturb the roots!

That's very true, and it means that you are exchanging one problem for another. So one of the most important things about tree nurseries is to find ways of temporarily **changing** the way the roots grow so that the root system is compact and can be planted without much damage. Then the roots can support vigorous stem and leaf growth (C 4, C 11-12, C 34).

Are there different kinds of tree nursery?

Yes, they can be:

(1) temporary or more permanent (C 21);

(2) of very different sizes (C 22), ranging from very small (dealing with about 10-100 young trees at a time) to very large (producing hundreds of thousands or millions of plants a year); ***and***

(3) used for a variety of different purposes, such as growing young trees for:

- (a) planting around the family home, garden and farm;
- (b) school and community planting (D 27 in Manual 4);
- (c) larger scale programmes by the government or private owners;
- (d) sale to other people; ***or***
- (e) use in experimental research (C 7).

What are the most important things about nurseries?

(A) Where they are put (C 20), particularly in relation to the planting site(s);

(B) Having reliable sources of clean water and of materials for making good soil mixes (C 6, C 23-24);

(C) Thinking carefully about the internal layout (C 22) before setting up the nursery (C 26);

(D) Producing favourable environments in them, so that the young trees have root systems that will give them a good chance of thriving (C 4, C 11, C 47);

(E) Protecting the young nursery trees against various kinds of stress, damage and loss (C 25, C 40-48); ***and***

(F) Building up a team of people (C 52) who understand the running of the nursery (C 50), the importance of regular watering (C 43) and care of the trees (C 40).

Will it be rather expensive?

Not necessarily - smaller nurseries at home or a school need not cost much, because:

(1) most of the materials required may be cheap or free (C 6, C 33), and can usually be obtained locally; ***and***

(2) some of the work can be done part-time, or by people who are prepared to work without cash payment because they will soon benefit from the tree planting (C 1).

Larger, permanent nurseries involve more expenditure, for instance to employ and train staff and workers (C 52) and install a water supply (C 24). Other costs could be involved in putting up buildings or making roads, and buying equipment or a greenhouse (C 48) for detailed research.

What sort of nursery problems could I face?

Most nursery problems are fairly easy to understand, foresee and correct (C 3, C 60; and A 2 and A 60-61 in Manual 1).
Some of the publications listed in sheets C 61-62 might help you avoid or cope with difficulties, and so could various useful contacts (C 53).

Are there some general guide-lines?

(A) The choice of which species to propagate depends strongly on the conditions in the planting site (D 20-29 in Manual 4), as well as the desired products and other benefits (D 30-42).
(B) Each young tree could grow into a large one, but to have a good chance of doing so it needs help, especially at key stages in its development (C 3, C 40).
(C) Care during these early stages does not mean that afterwards it will be unable to stand up to climatic stresses such as drought (C 41, C 47).
(D) Besides difficulties of a biological kind, think about possible problems involving people's differing views (D 4-5 in Manual 4), changing markets, breakdown in the systems for checks and care (C 40, C 66), and the risk of accidents (C 3; and D 66).

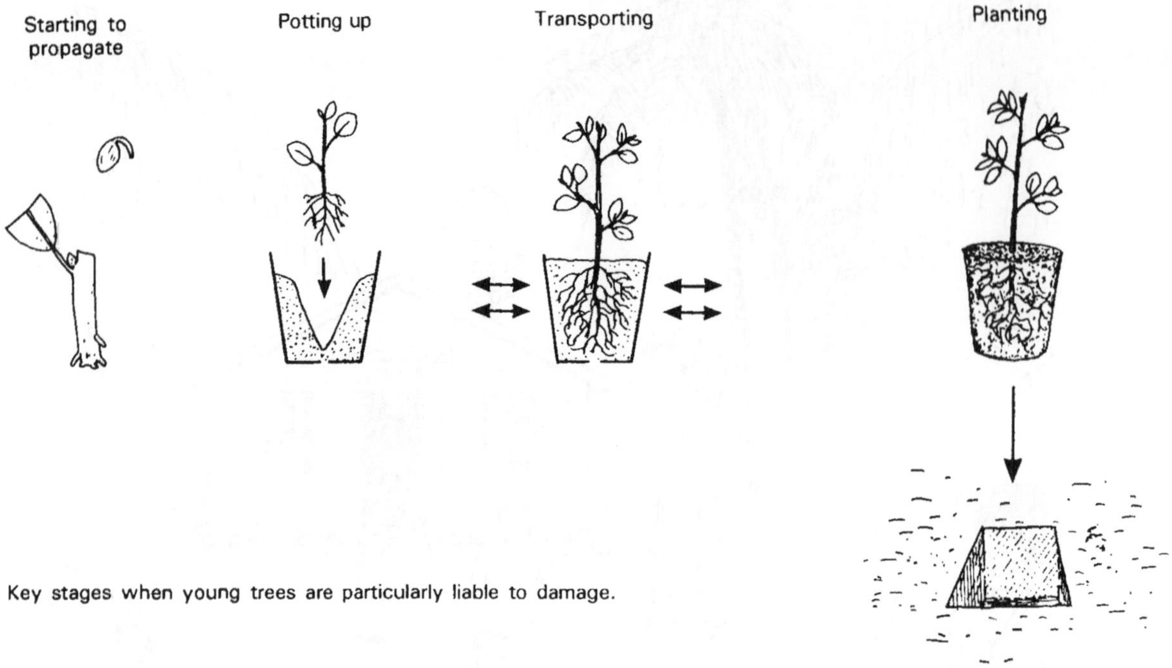

Key stages when young trees are particularly liable to damage.

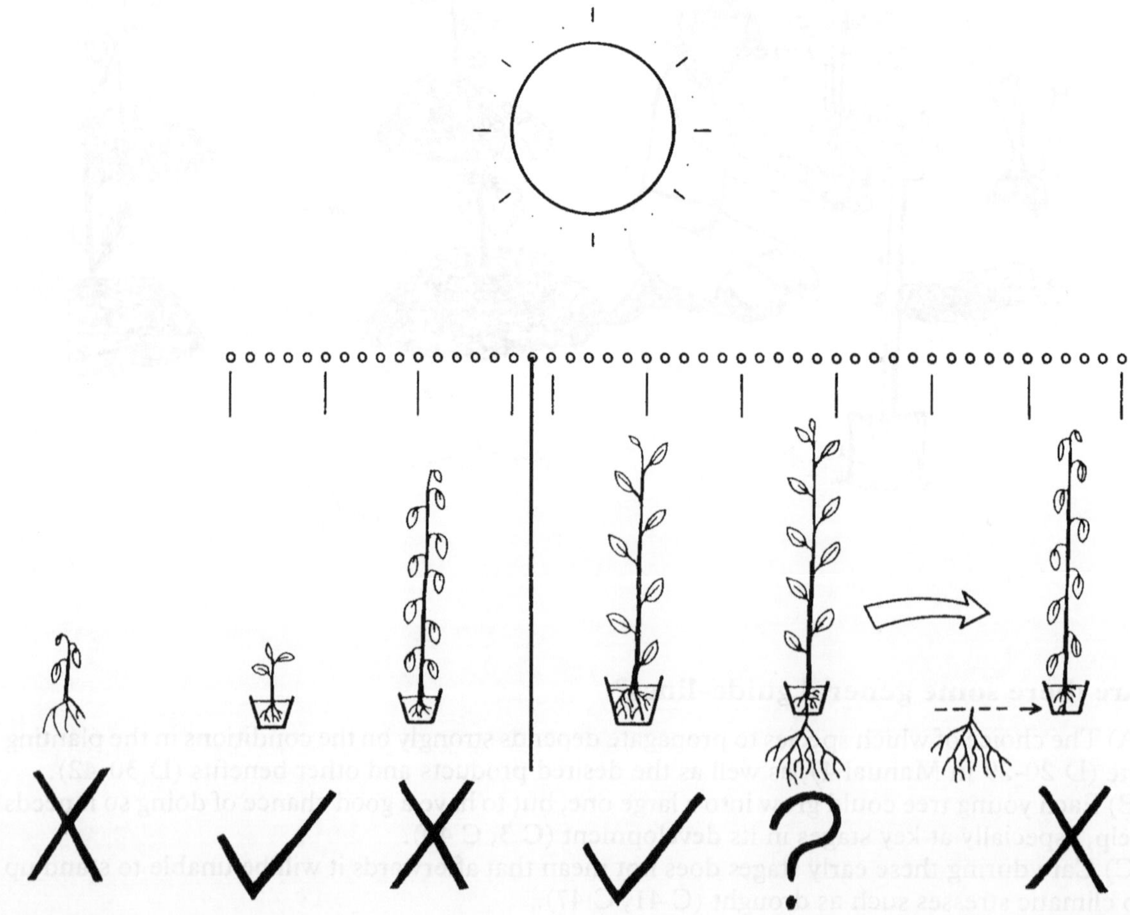

Importance of maintaining the water balance.

Introduction C 3
- *overcoming problems when growing trees*

Are there a lot of nursery problems?

Yes, if you make a list of all the things that could go wrong; ***but***
No, in the sense that most of them can easily be avoided.

When are young trees most at risk?

(A) When they are very small;
(B) If watering is neglected or done badly (C 41, C 43);
(C) Whenever their root systems have been disturbed (C 40); ***and***
(D) If their shoots are repeatedly damaged by pests or larger animals (C 45-46).

What does that mean in practice for nursery problems?

Stages at which young trees are most susceptible to damage include when:

(a) seeds are just germinating (Manual 2);
(b) cuttings have not yet formed roots (A 40-45 in Manual 1);
(c) seedlings and newly rooted cuttings are being potted up or transplanted (C 42);
(d) plants are being prepared for transporting to the planting site (C 47).

What about after they leave the tree nursery?

Young trees are particularly at risk:

(1) during transport and before planting (Manual 5);
(2) especially for the first few days after planting, and during the next 2-4 weeks.

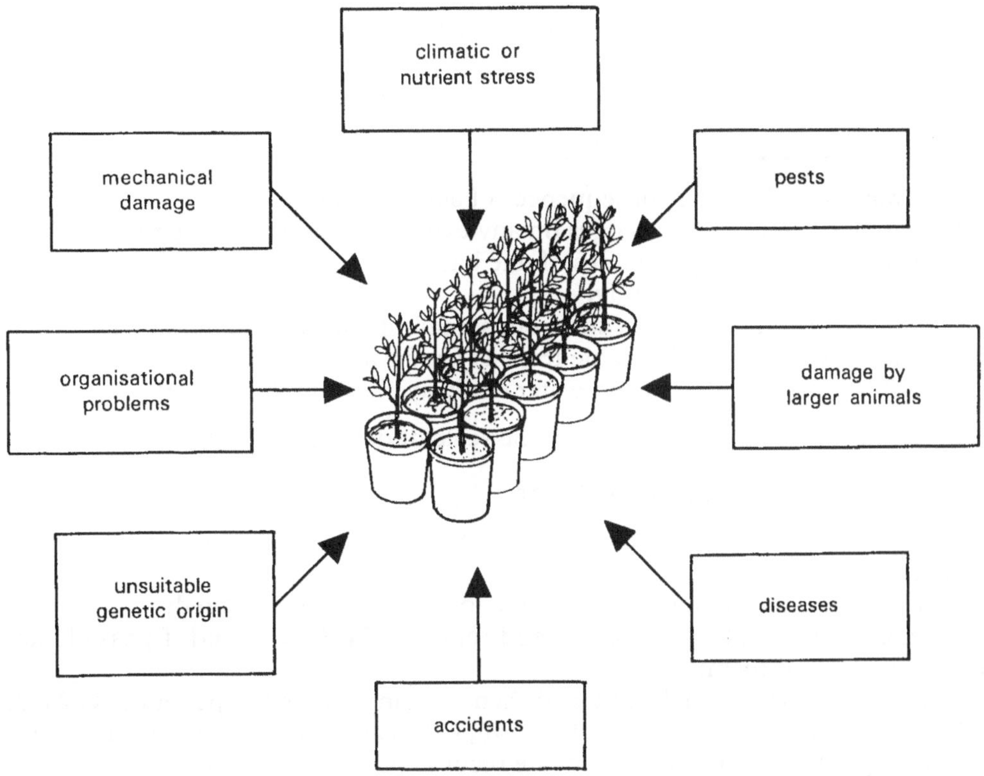

And how about before nursery work starts?

Yes, success also depends on decisions taken beforehand, such as:

(a) choosing appropriate tree species for the site and the expected usefulness (D 30-42 in Manual 4);
(b) using suitable genetic origins (C 5 and Manuals 1-2);
(c) planning the way seeds are to be handled and stockplants managed; ***and***
(d) arranging for frequent, regular visits to the nursery.

But aren't there bound to be problems?

Yes, but for instance you could:

(A) think about potential snags that might come up in your situation;
(B) make frequent checks (C 40), and keep on the lookout for signs of trouble (C 60);
(C) try to reduce the likelihood of problems occurring (C 50; and D 4 in Manual 4); ***and***
(D) make plans to lessen the impact of those that do happen.

What sort of things might go wrong?

You could think about three different kinds of problem (A 2, A 61 in Manual 1):

(1) **accidental hazards,** such as drought, flooding, storms or fire (C 25, C 40-41, C 46);
(2) **organisational difficulties,** including a lack of information, experience or skills; or a shortage of finance, manpower, tools or materials; ***and***
(3) **biological problems,** as for example:
- (a) difficulties with seed germination (Manual 2);
- (b) poor rooting of cuttings (A 50 in Manual 1);
- (c) death or unsatisfactory growth of young trees (C 60);
- (d) competition from other plants (C 44);
- (e) damage from pests or diseases (C 45); ***or***
- (f) eating and breaking by larger animals (C 46; and D 14-15, D 34 in Manual 4).

Is there anything I can do to prevent accidents?

Yes there is, though the risks will always remain, whatever one does. However, one could lessen the chance of accidents happening for instance by:

(A) careful selection of the nursery site (C 20);
(B) maintaining tools (C 50) and fences (C 46) well, and any buildings and vehicles;
(C) removing inflammable materials, and digging drains or building barriers against floodwater (C 23, C 25; and D 65 in Manual 4) ***and***
(D) learning about various kinds of risks, training people in safety measures (C 50, C 52, C 54; and D 66 in Manual 4), and not leaving the nursery unattended for long periods.

What about reducing the losses?

Examples of ways to reduce the consequences of accidents are:

(1) planting shelter belts or hedges on any exposed sides of the tree nursery (C 25);
(2) not keeping all the young trees of one batch in the same place (C 65);
(3) keeping a first aid box handy;
(4) stopping fires that start outside from spreading far into the nursery by:
- (a) growing strips of suitable trees as *fire-breaks;* ***and***
- (b) having fire brooms and water easily available in the nursery.

Items (A-D) above may also help to lessen the scale of damage.

How about problems of organisation?

Some of these are discussed on sheets D 4-5 in Manual 4. Hints that are particularly relevant to tree nurseries include:

(a) making sure the land will be available for as long as it is needed;
(b) using locally available materials and tools (C 24, C 51), and if possible keeping a sufficient stock of them;
(c) starting on a fairly small scale and then building on your experience (C 21-22);
(d) spending enough time on checks (C 40, C 66) and training the nursery team (C 52);
(e) making plans and keeping adequate records (C 54, C 64-66).

Supposing my plants don't grow well?

Amongst problems that might be involved are:

(A) an unsuitable **tree species** for the climate of the area (C 10-15; and D 10-11 in Manual 4);

(B) an inappropriate **genetic origin** for the region or the intended purpose (C 5);

(C) unfavourable **potting mixtures** (C 6) or **nursery soil** conditions (C 23; and D 12-13 in Manual 4);

(D) poor **watering regimes** (C 13, C 43);

(E) lack of important **nutrients** and/or **micro-organisms** (C 30-34);

(F) **bud** or **seed *dormancy*** delaying growth or germination (C 12; and Manual 2);

(G) damage by **weeds** (C 44; and D 14 in Manual 4), larger **animals** (C 40, C 46; and D 15 in Manual 4) or **pests** and **diseases** (C 45; and A 52 in Manual 1).

This all sounds rather difficult!

Well, in practice, one seldom has to deal with more than a few problems at a time!
Sheet C 60 summarises various symptoms of poor nursery growth and their likely causes. *(For poor rooting of cuttings see A 50 and A 61 in Manual 1. For germination problems see Manual 2.)*

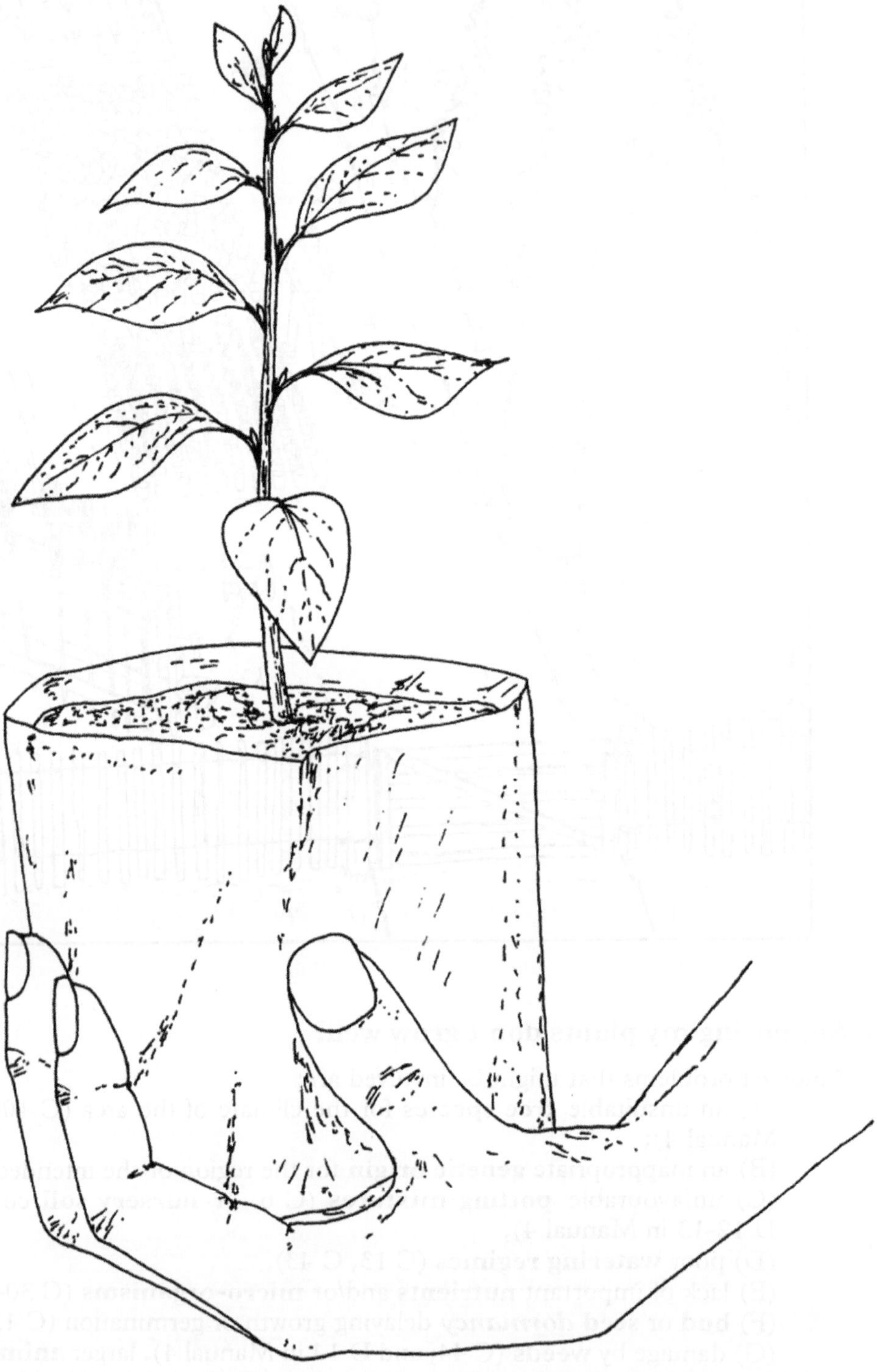

Introduction C 4
- *what makes a good tree for planting?*

What are the features of good planting stock?

The young trees should be:

(A) of appropriate **species** for the particular planting site (D 20-29 in Manual 4);
(B) from suitable **genetic origins** (C 5) for the intended purposes (D 30-42); ***and***
(C) in a favourable **physiological condition** (C 10-15) for being planted (C 47).

Isn't it really just the species that matters?

No, the other two points are also important, because:

(1) a lot of losses and unnecessary problems can be avoided; ***and***
(2) good planting stock is more likely to establish quickly and grow well, and to give increased benefits and yields.

Is it difficult to grow good nursery trees?

No; it mostly involves setting up and following straightforward procedures (C 50-54, C 63-66), and gaining experience of dealing with problems (C 3, C 60).

Improvements in approaches and techniques can also come through training courses (C 50), local contacts (C 53), publications (C 61-62) and by doing informal or formal experiments (C 15; and D 6 in Manual 4).

Which tree species might I choose?

Those which are likely to:

(A) thrive in the average climate experienced in the planting site (D 11, D 30 in Manual 4), and also survive extreme conditions that could occasionally occur;
(B) fit in with the existing patterns of vegetation and land use (D 14, D 50-54), and preferably improve the soil (D 32); ***and***
(C) provide the desired products and benefits (D 30-42).

Should I concentrate on just one kind of tree?

Occasionally yes, for example in specific situations, such as:

(1) establishing shelter belts in dry areas (D 41 in Manual 4);
(2) restoring patches of mangrove woodland (D 26); ***or***
(3) reclaiming degraded farmland (D 21-22); ***but***

Generally no, because:

(a) mixtures of trees usually provide better soil protection and maintenance of fertility than pure stands (C 30; and D 30, D 53 in Manual 4);
(b) even on a small piece of land there can be a variety of conditions, suitable for different tree species (D 12, D 20-29);
(c) you may well want several benefits and products, and prefer to have yields spread out over time; ***and***
(d) Because unexpected things may happen, it is risky to "put all your eggs in one basket". Species diversity acts as an 'insurance' against market changes as well as possible heavy losses from unusual weather conditions, pests or diseases.

Don't natural stands have hundreds of different species?

This is often true of tropical forests, and most of them can give useful products. However, when starting a tree nursery, you can choose a selection of several local and introduced trees (D 31 in Manual 4):

To provide young trees for your farm, you might begin with perhaps 4-6 species for fruits, fodder, shade and marking boundaries.

For a medium-sized nursery, you could decide on 5-10 main species and 3-8 grown on a smaller scale, plus a few new kinds for trying out.

For nurseries selling young trees, and those in parks and Botanical Gardens (D 28, D 41 in Manual 4), many more species of tree and shrub will usually be propagated.

Does this Manual give specific instructions for growing the major plantation trees?

No, it aims to present **the general principles upon which** trees of the humid and semi-arid tropics can be raised as good planting stock.
Like the other Manuals in the series, it may encourage the writing of local sheets on the many other valuable tree species that have often been neglected in favour of a few plantation crop trees.

Why do different genetic origins matter?

Individual trees generally vary so much within a species that it would be a big advantage to use those which are most suitable for the purposes in hand (C 5).
While these are being chosen and tested, one can still make considerable progress by excluding inferior types.

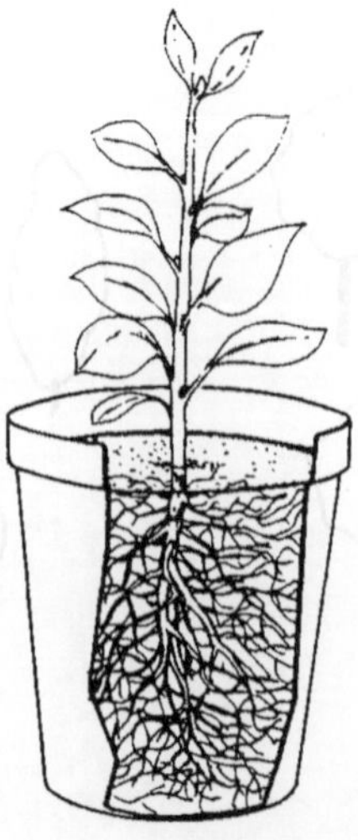

What is "a favourable physiological condition for being planted"?

This term includes many points (C 10-15) which together give the tree a good chance of becoming established, tapping the available soil moisture and nutrients, and growing well. Amongst the most important are that each tree should:

(A) have a **good, bushy root system** (C 6, C 11) that is not *pot-bound,* but is compact enough to be planted without too much damage;
(B) have a **moderate-sized shoot system** (C 34) that can, without dying back, produce new growth rapidly once the roots have recovered from planting shock;
(C) be free of **pests** and **diseases** (C 45); ***and***
(D) have been suitably **'hardened'** (C 47) and recently **watered** (C 43), so that it can tolerate moderate climatic and mechanical stress (C 13, C 41) in transit.

Shouldn't trees be treated roughly for them to survive planting out?

No, this is a common but mistaken idea. The basic function of a nursery (C 2) is to:

(1) raise seedlings and root leafy cuttings under **favourable** conditions in a protected environment (Manuals 1-2);
(2) encourage good growth to the size required (C 34); ***and then***
(3) harden the young trees off gradually (C 47) before they go to the planting site.

How important is the root system?

It is the main key to success (C 11, C 34). A desirable root system can generally be obtained by:

(a) using suitable containers and an effective potting mix (C 6);
(b) watering carefully and regularly (C 43);
(b) minimising other kinds of stress (C 41); ***and***
(d) pruning back longer roots to encourage many branch roots to form ***inside*** the pot (C 6, C 11).

Alternatively, young trees can be grown in good soil in a nursery bed with more frequent root pruning. The trees are then taken for planting complete with the soil block around the roots.

Isn't a strong tap root best?

No, only with direct planting (C 2), and for occasional species where it is needed. Contrary to another common misconception, they are generally unsuitable for nursery grown trees that will be planted out later on, because:

(A) **tap roots** often occur singly, have few branch roots at an early stage, can get twisted round and round inside a pot, and are difficult to plant intact; ***whereas***
(B) **branch roots** can occur in larger numbers, usually produce many side branch roots rapidly, are less prone to become set in a coil, and are easier to plant undamaged.

Branch roots can also be inoculated with useful micro-organisms (C 30-32).

What about the shoots?

Rapid early growth in the nursery is desirable, but later the rate of shoot growth should slow down or even stop temporarily (C 12, C 34). This is because a sturdy stem and moderate area of dark green leaves mean less water stress than a tall plant with a lot of large, paler-coloured leaves (C 41). In order to maintain water balance after planting (C 13), the roots first need to have time to grow and branch into the soil before strong shoot growth is made.
However, vigorous growth in the nursery is not a problem if striplings or stumps are to be planted (C 47).

In general, the tree's capacity to produce vigorous new shoots **after** planting is more important than its starting height.

Are pests and diseases very important?

Most of them can be avoided or dealt with by simple methods (C 45), but there are a few that can do a lot of damage. It is a good idea to learn how to spot early stages of an attack, and to check the young trees frequently (C 40, C 66).

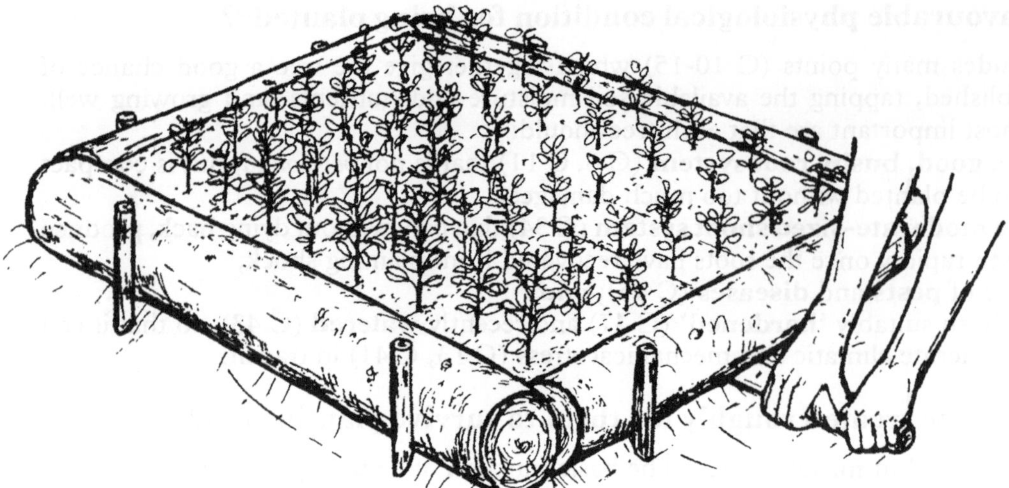

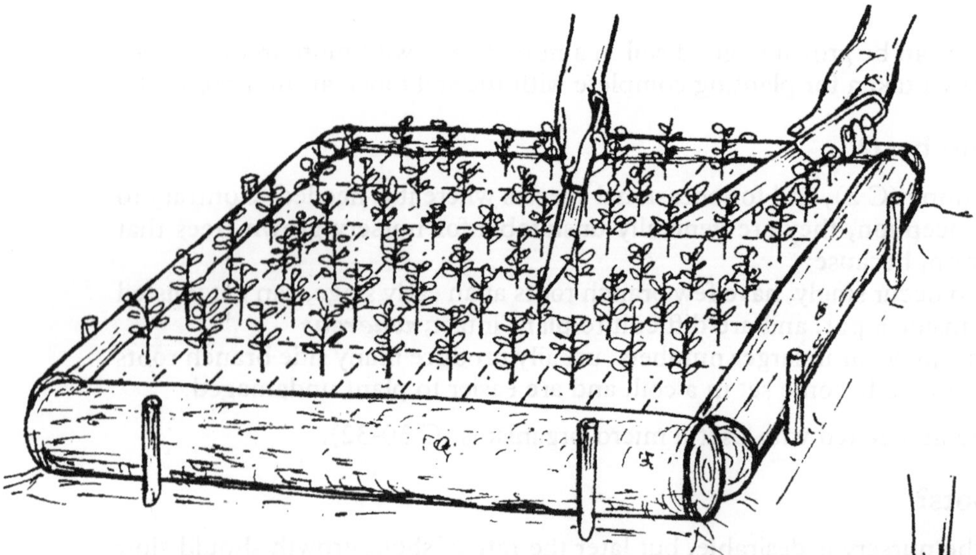

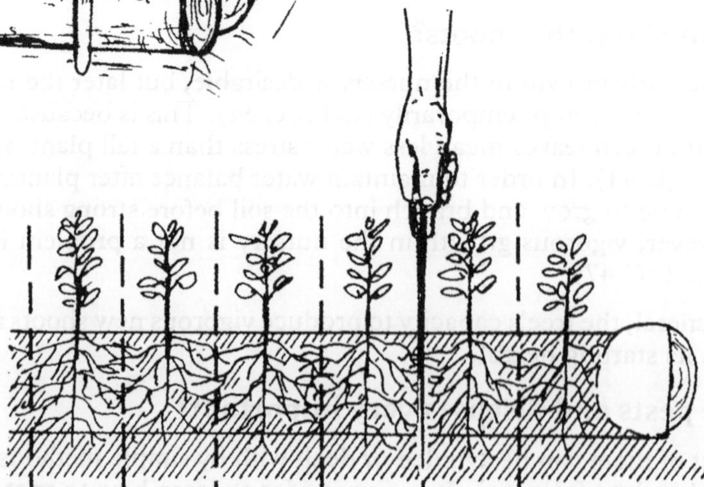

Regular cutting of the longer roots promotes good bushy root systems in the 'root-pruned soil block' method of growing young trees.

Is more research needed on growing good trees?

Yes it is (C 7). A fair amount is known about the underlying principles of tree growth (C 10-15, C 62-A) and general methods for tree nurseries (C 61), but much less work has been done on:

(1) which factors particularly affect the growth of individual tree species;
(2) how these may indicate the need for different nursery practice;
(3) achieving the most useful close associations with micro-organisms (C 30-32); ***or***
(4) stimulating flowering in the desired parent trees (Manual 2); ***and***
(5) studying variations in the underlying processes within the cell (C 10, C 14).

Why haven't most local trees been looked into?

Because:

(a) many of these species used to come up by themselves (C 1);
(b) most research effort and funds were directed at cereals and plantation crops; ***and***
(c) hardly anyone had thought about doing it.

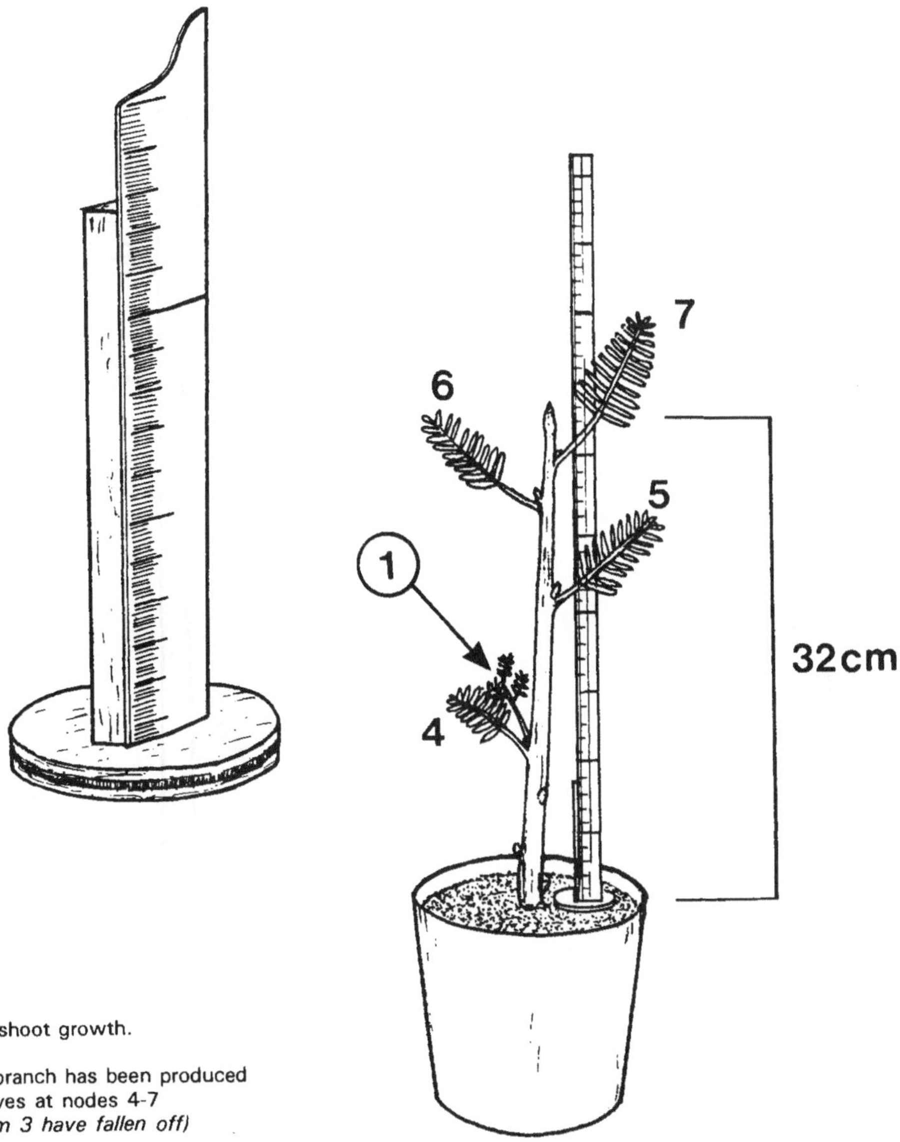

Assessing shoot growth.

1 – one branch has been produced
4-7 – leaves at nodes 4-7
(the bottom 3 have fallen off)

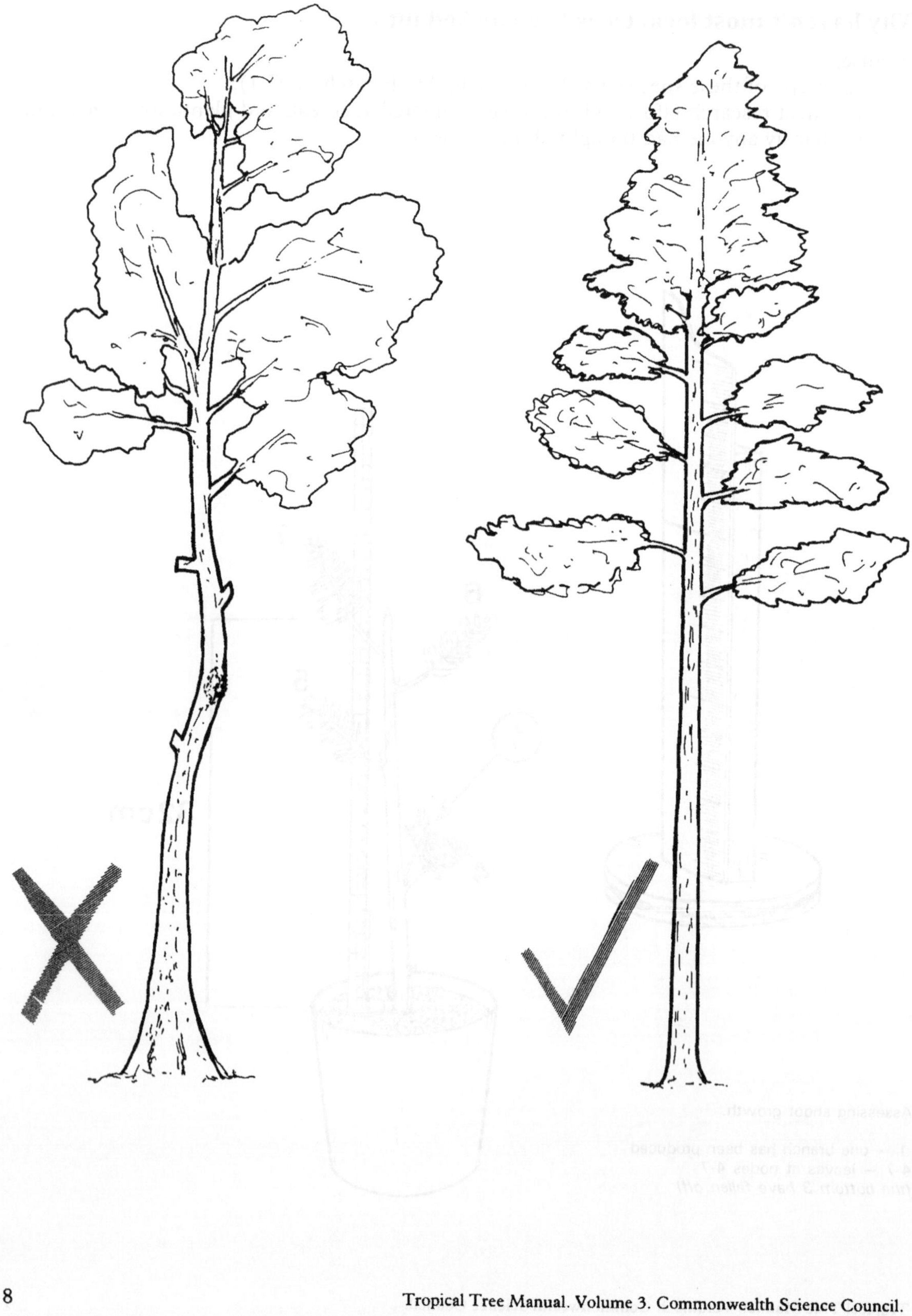

Introduction C 5
- *'domestication' of trees*

What is 'domestication'?

Improving a tree species that has grown only in the wild, by finding or producing particular varieties, populations or clones that are more suited to the local environment and the purposes for which the trees are to be planted.

Is that important?

Yes it matters quite a lot, because:

(A) big differences often exist between individual trees within a 'wild' species;
(B) this variability is generally found in important features such as height and diameter growth, stem form and branching habit, fruiting characteristics and tolerance to disease;
(C) such features are usually strongly influenced by genetic characters inherited from the male and female parent trees.

The features of trees also depend on the environment they have grown in, internal changes within them, and chance occurrences (A 10 in Manual 1).

Isn't it a good thing to keep a lot of genetic diversity?

Yes, in the sense of:

(a) not clearing most of the sites where a tree species grows naturally;
(b) not cutting down all the best trees; ***and***
(c) not planting closely related seedlings or a single clone in large blocks; ***but***

No, for the practical grower, because:

(1) all the available diversity need not go into each planting (C 62-E), provided some is conserved for the future in *gene banks* and natural stands (Manual 2); ***and***
(2) a domesticated stand can still contain plenty of genetic variability to protect it against the risks of over-uniformity.

Genetically uniform planting stock is very useful for experiments (C 7).

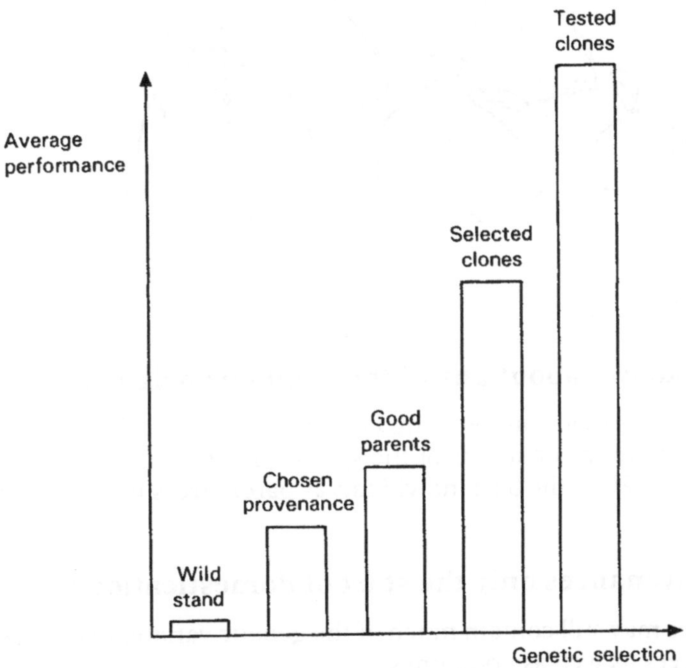

How can I improve the genetic quality of my nursery trees?

The first steps in domestication can be taken fairly easily, for instance by:

(a) choosing a suitable *provenance* (part of the natural range) to use (Manual 2);
(b) taking seeds from a sizeable group of healthy trees with favourable characteristics; ***and***
(c) rooting cuttings of selected clones (A 11 in Manual 1).

Will that give me the highest yields?

An improvement over unselected material is likely, but definitely not the full potential at these early stages of the process. Moreover:

(1) if the *selection* has been based only on observing the features of the parent trees or stockplants, testing in field trials (D 29 in Manual 4) will be needed to confirm whether and to what extent they are *genetically* superior; ***and***
(2) yields tend to be influenced by more than one genetic characteristic.

How much difference can selecting the provenance make?

Quite a lot, especially if the species has an extensive natural range. One provenance may do very poorly or die, while others thrive.
In other cases, differences can be much smaller, or hard to detect.

So how do I choose the best provenances for seed collection?

For local trees (D 31 in Manual 4):

(a) where possible, use **natural stands in the locality** as seed sources;
(b) if not, collect seed from good **plantations made with local seed.**

For introduced species:

(1) preferably use seed collected in the area **where the species grows naturally,** of provenances that have done well in trials in your region;
(2) alternatively collect seed **from trees thriving locally,** but avoid using only one or two trees or a small group that might be too closely related to one another.

For a provenance trial, make sure you record the origins of the various seed lots, label them and grow separated batches of seedlings under uniform conditions (C 7, C 54, C 64-65).

Suppose nothing is known about any of the available origins?

It is still worth selecting your seed sources. You could:

(a) look for thriving groups or stands of the species; ***and***
(b) choose at least 10-20 of the best individuals as parent trees for pollen production and seed collection.

Why is selecting provenances only the start of domestication?

Because your young trees may still contain much of the genetic variability of the species. There will be *inherently* poor trees as well as good ones.

Is that why one collects seeds from the best parent trees?

Yes, this is the next step, especially if you can make a *seed stand* (Manual 2), where:

(1) the inferior trees have been removed; ***and***

(2) there are plenty of above-average trees.

However, the potential for further domestication will generally still be large.

Why is this so?

There are several reasons, including the following:

(1) one seldom knows where the *pollen* has come from, so *mother-tree selection* covers only half the inherited characteristics;

(2) the genetic features of the two parents are not all passed on to the young trees, and become considerably mixed during flowering and seed formation;

(3) if the parent trees are closely related, their *progeny* (the seedlings resulting from crossing them together) could show *inbreeding depression* (Manual 2);

(4) when a few of the trees in a stand produce a lot more pollen and/or seed than the others, the seed you collect may not reflect the features of the stand as a whole, and might well have a greater tendency to flower than their parents; ***and***

(5) the best trees may not produce much pollen or seeds, so may not be represented.

How can one get over these problems?

(A) By **open-pollinated progeny trials** to compare the seed from individual mother trees;
(B) By **controlled breeding** (crossing) of parent trees, followed by **full progeny trials** to test the seed originating from a particular male or female parent tree, or from specific combinations of parents; ***and***
(C) By direct **vegetative propagation** of selected trees.

What are the advantages of progeny testing?

(1) It can take domestication further than provenance selection, particularly when both parents are known, and they flower or can be stimulated to do so (Manual 2);
(2) It may allow *hybrids* to be bred between related tree species or between provenances of the same species. These may show considerable *hybrid vigour* and can sometimes combine desirable features of each parent.

How about the disadvantages?

There are several of these, particularly with controlled pollination, including:

(a) having to wait many years, and then climb high up in a large tree to reach the flowers, and again later to collect the ripe fruits, because young trees and shaded lower branches are usually not reproductive (Manual 2);

(b) a lack of flowering on the chosen parent trees, flowering times that do not coincide, or *incompatibility* between them;

(c) some crosses not producing enough viable seed for the progeny trial;

(d) uncertainty about whether early performance of a progeny is necessarily linked to how it will grow later on; ***and***

(e) difficulties in obtaining adequate and regular supplies of improved seed after the trial has been done.

But all this sounds much too difficult!

Yes it is really only feasible for large research organisations that can plan a long-term breeding programme. However, some success has been achieved with pines and other tree species that start regular flowering early in life.

Can't one get improved seed from the progeny trial?

No, it is usually **unsuitable** as a seed source, since the characteristics of a second generation of progeny could be quite different.

What has to be done, then?

Following the progeny trial, *mature* clones are made by grafting shoots from the crowns of the best parent trees on to seedlings (A 3 in Manual 1). Several of these are then planted in a *seed orchard* (Manual 2), with the aim of producing quantities of improved seed.

Wouldn't it be better to use rooted cuttings as planting stock?

Yes, once suitable provenances are known, vegetative propagation of *juvenile* shoots often offers quicker, cheaper and more reliable domestication.

For instance, taking coppice shoots and rooting them as cuttings (A 21 in Manual 1) offers the double opportunity of:

(A) growing the 'same' trees over again; ***and***

(B) multiplying them into many new trees, each with the same genetic potential as the original ones.

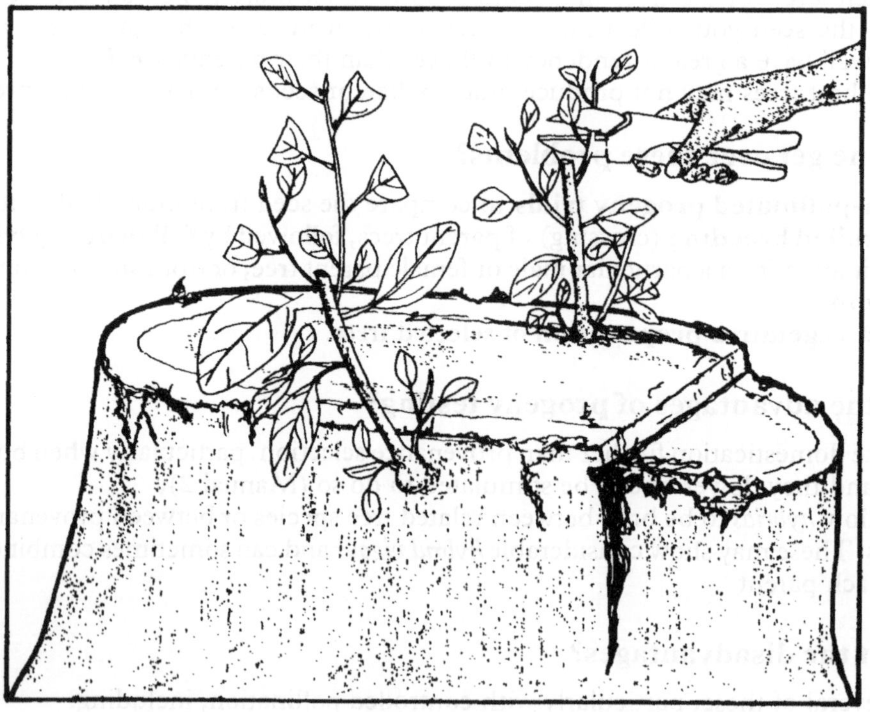

How else could I get a supply of cuttings?

Suitable juvenile shoots can also be taken from:

(1) young seedlings, preferably those that were found to be superior in a selection test (A 13 in Manual 1);

(2) stockplants that are managed to provide plenty of suitable shoots (A 20-27); ***and sometimes***

(3) plantlets from micropropagation (A 5).

Isn't this a much more direct way of selecting trees?

Yes it is, because:

(a) one is using clonal sets of young trees with the **same genetic potential** as the selected trees, rather than seedlings that are all different from their parents; ***and so***

(b) clonal trials are a **more precise check** on domestication than progeny tests, because the 'same' tree can occur many times over, whereas each seedling in a given progeny is different.

Vegetative propagation is the way farmers and scientists have selected crops like banana, cassava, mangoes, tea, coffee and rubber.

Aren't there some disadvantages with vegetative propagation?

There are a few; for example that:

(1) a good tree has to be cut down in order to start a clone from coppice shoots; ***and***

(2) rooting may prove difficult in a small proportion of tree species.

Note: there are considerable risks in planting stands that consist of only one or two clones (A 11 in Manual 1). Between 10 and 30 clones is advisable.

But can rooted cuttings grow into tall trees?

Yes they normally do, as long as they:

(a) are of a species and clone with the potential to grow tall;

(b) originate from juvenile stockplants (A 6 in Manual 1); ***and***

(c) have formed several roots, rather than only one (A 40).

Then why aren't they utilised more often?

They should be! Although widely used in agriculture and horticulture, this method of domestication has been neglected for most other kinds of trees, perhaps because:

(1) most have a longer life-cycle than farm crops and ornamental shrubs;

(2) by the time you know you have a good tree, its crown shoots usually do not root;

(3) if these are grafted on to a seedling the plants would be expensive, and the adult growth habit could be unsuitable for growing tall trees; ***and***

(4) there were several mistaken assumptions about vegetative propagation (A 3 in Manual 1).

But won't domestication be too expensive to bother about?

No, because:

(A) cuttings can usually be rooted very effectively without costly equipment (A 30-31 in Manual 1);

(B) using undomesticated planting stock generally means inferior trees; ***and***

(C) genetic improvement continues to provide benefits into the future.

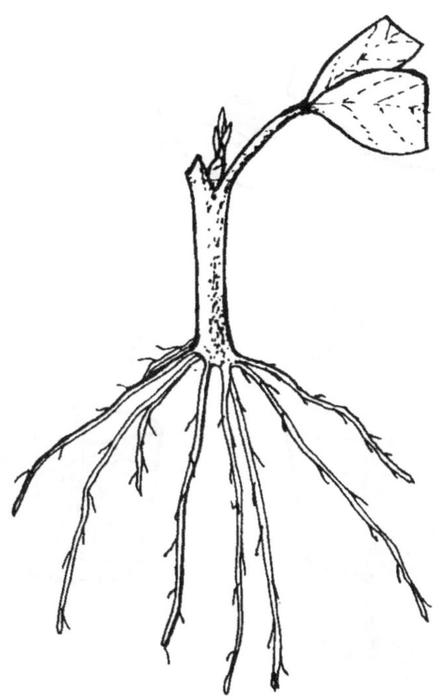

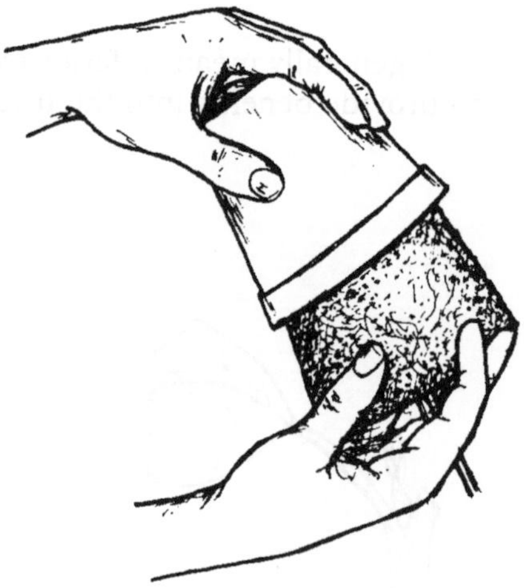

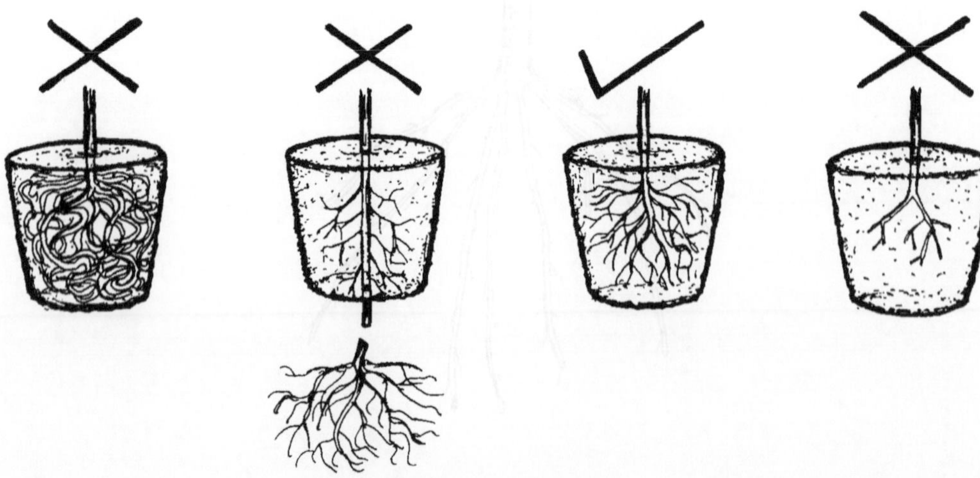

Introduction C 6
- *containers and potting mixes*

Will I need containers for my trees?

Yes, in many circumstances. Transplanting young trees directly from a nursery bed:

(A) is easy only when they are still too small for planting in the field; ***and***

(B) usually disturbs the root systems too much when they are big enough to plant out; ***but***

No, if longer roots are regularly pruned back to stimulate plenty of branch roots, and the young trees are then taken to the planting site complete with a block of soil (C 4); or if stumps or striplings are to be planted (C 47).

Why wouldn't very small plants succeed in the field?

Because they would usually:

(1) be badly damaged in transit (Manual 5);

(2) have too small a root system (C 4, C 11) to support the young tree in the harsher environment of the planting site;

(3) be quickly overgrown, even by low-growing weeds;

(4) have a low chance of surviving if parts were eaten by animals; ***and***

(5) be difficult to find during weeding, and therefore easily cut off.

Aren't larger plants sometimes used as bare-rooted planting stock?

Yes they are. However, the exposed root systems of young trees can lose quite a lot of water within a minute or two, in humid as well as in drier parts of the tropics (C 13). So for this method to be successful:

(a) the trees leaving the nursery need to have a very good root system (C 34);

(b) their *leaf area* generally has to be reduced, so that there is less water loss (C 12-13);

(c) the roots have to be covered with wet leaves, cloth or polythene sheeting, and preferably transported at night.

Bare-rooted plants are sometimes moved for planting during the cooler season in the temperate zones, sub-tropics and tropical uplands, when *climatic stress* is less (C 41).

How do containers help?

(A) they can hold enough good soil to allow small plants to grow large enough root systems for successful planting (C 4, C 34);

(B) the young trees can be moved around without much damage to the roots, and then the whole *root ball* removed from the pot at planting time. (*Biodegradable* pots are not removed before planting, but gradually rot away, so that root disturbance is minimal.)

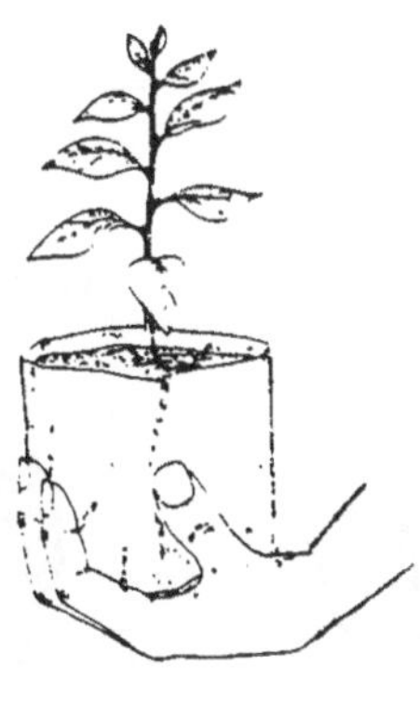

What are the features of a good container?

(1) Permitting good root and shoot growth by allowing excess water to drain out;
(2) Being strong enough to last a reasonable time, but light in weight;
(5) Being cheap and readily available; ***and***
(3) Releasing the root ball easily at planting time.

What can pots be made of?

They are manufactured out of several different materials, including:

(a) fired or unfired clay;
(b) polythene sheeting;
(c) many other kinds of thicker plastics;
(d) various biodegradable substances, such as compressed paper or peat.

But what if I can't afford to buy pots?

This need not matter, as serviceable pots can be made out of such things as:

(e) sections of bamboo stems, with a hole made through the cross wall for drainage;
(f) small baskets, made for instance out of woven raffia, split cane or bamboo;
(g) strong leaves, such as *Bassia, Dyospyros* or banana, which can be shaped and tied around a bottle that is subsequently removed;
(h) suitable soil, moulded into shape with a press;
(i) waste veneer (C 33) rolled into a tube and stapled;
(j) small pieces of waste wood nailed together;
(k) used cardboard milk or fruit juice cartons;
(l) the lower parts of used plastic bottles or recycled polythene bags; ***or***
(m) old cans, preferably sterilised in a fire.

Stages in making pots from strong leaves.

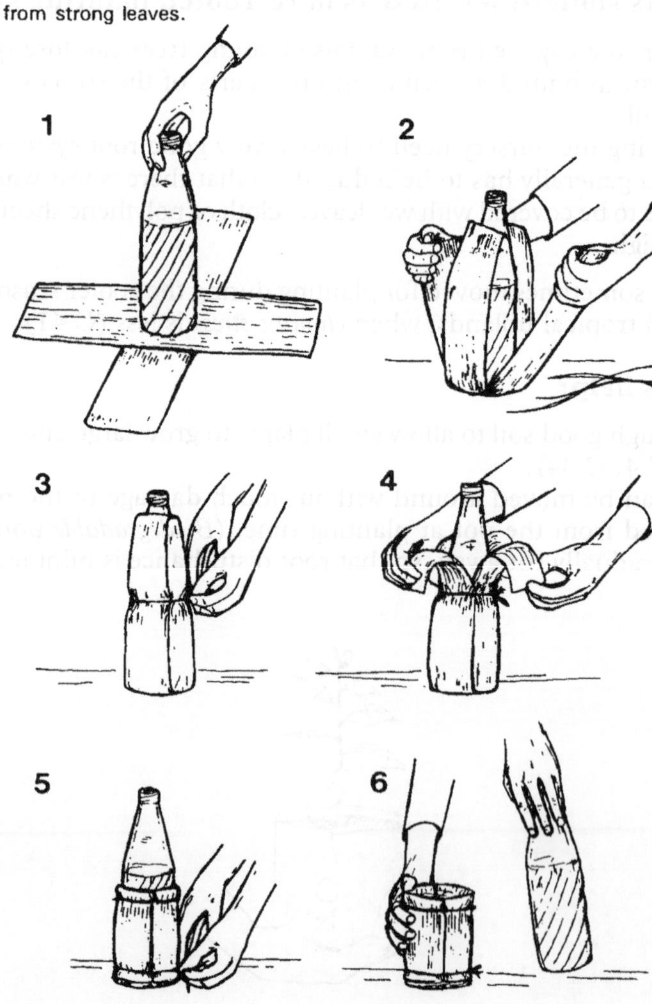

Are open cylinders any good?

Yes, if they are either:

(A) stacked together in a shallow box; ***or***

(B) given a base by adding about 3 cm of soil containing silt or clay, and tamping this down with a stick to compact it. (Don't forget to make drainage holes!)

Which kinds of pot are most commonly used?

This differs a lot from place to place, and depends on the size of the nursery (C 22). A common type is a thin bag made out of polythene sheeting (black, white or clear), with some small drainage holes punched in it.

What are its advantages?

These poly-bags are very light in weight and quite strong. They can be made in various sizes, manufactured in quantity, and are relatively cheap and easy to transport.

Do they have some disadvantages?

(1) Since the poly-bag has to be torn to release the root ball, it cannot be used again;
(2) Because the material bends easily, the root ball is more liable to be damaged in the nursery and in transit, and can break apart during planting;
(3) As with many other containers, the roots may go round and round the bottom, producing poor planting stock (C 11, C 34, C 47); ***and***
(4) If the drainage holes are in the bottom of the pot, *rooting through* into the soil beneath can easily occur. Such trees may grow well in the nursery, but when they are moved and the roots broken they quickly wilt, and often die (C 13, C 41).

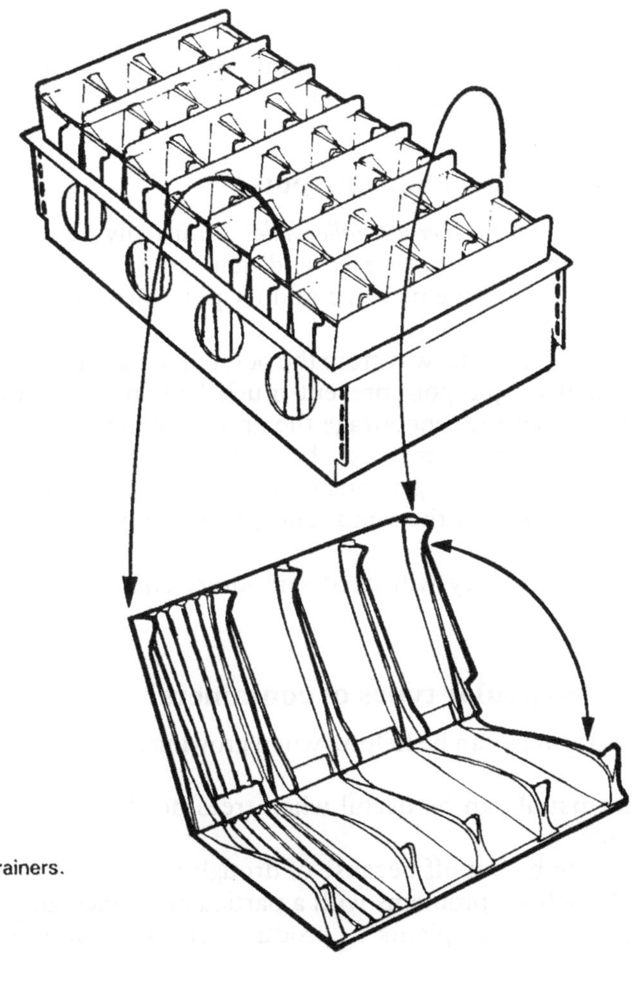

Example of root-trainers.

Can such difficulties with the root system be avoided?

Not altogether, since one is restricting it to a confined space (C 2). However, the problems can be reduced, for instance by:

(a) discouraging the roots from emerging by putting the containers on flat stones, concrete or polythene sheeting (without having standing water on it), or by keeping the containers clear of the ground on wire mesh;
(b) moving the pots regularly, to break off young emerging roots before they grow long, or by using another method of root pruning;
(c) not leaving young trees too long in very small containers, but potting them on (C 42) into somewhat larger pots;
(d) using 'Root-trainers', various types of containers that have vertical grooves and ridges that encourage the roots to grow downwards rather than in circles (C 61-B).

Outgrowth of roots can also be restricted and branching promoted if the inside of the pot is treated with *copper carbonate*. However, although copper is a micro-nutrient (C 14), it is toxic to plants except at very low concentrations. This technique should therefore only be tried by a well-trained nursery team (C 52), taking care not to pollute the environment (D 16 in Manual 4).

How big should the pots be?

Not too small, so that:

(1) there will be a reasonable *rooting space* and enough reserves of nutrients (C 14-15, C 33) to produce good planting stock;
(2) the pots do not dry up too quickly (C 13, C 43); ***but***

Not too large, resulting in:

(a) root systems that are too big to be planted well (C 34 and Manual 5);
(b) shoot systems that are liable to die back (C 41);
(c) a lot of unnecessary weight to carry to the planting area.

For most purposes, the volume of soil in a container should not be less than 250 cm^3 or more than 2500 cm^3 (0.25 to 2.5 litres - see C 63-D).

What else is important when choosing pots?

(1) **Drainage holes:** types with *several smaller holes* are usually better. A single hole may get blocked, making the potting soil waterlogged and damaging or killing the young tree (C 11, C 13). If rooting through is a problem, you could try pots with small holes *near the base of the sides.*
(2) **Colour:** *light* colours reflect heat away from the pot, reducing the risk of the roots overheating (C 41). *Dark* colours and *opaque* pots prevent much light reaching the roots, but may get hotter. *Clear* containers sometimes encourage the growth of algae and mosses.
(3) **Shape**: *tapered* pots make it easier to get the root ball out intact, which may allow you to use them again; *square* pots utilise the growing space efficiently; and reasonably *wide* containers are less likely to fall over easily in the wind, and provide more space for the shoots (C 12, C 34).
If tall, narrow pots are used, they will need temporary supports and perhaps spacing out (C 42).

When might I need particular types of containers?

(A) **Standardised containers** can be helpful when growing a set of similar plants for research (C 7, C 15);
(B) **Larger pots than usual** can be useful when trees need to remain in the nursery for a longer time, for example:

(1) if planting had to be put off because of drought;
(2) when there have been problems with a particular species; ***or***
(3) for growing potted stockplants to produce repeated harvests of cuttings (A 23 in Manual 1).

Where should empty pots be stored?

Preferably inside a simple store (C 20), because:

(a) pots made out of leaves, wood, bamboo and other biodegradable materials need to be kept dry;

(b) most plastics become brittle after a time if left in the sun, and tins will rust away;

(c) most containers need protection from domesticated animals, while mice and other *rodents* may chew most kinds, even plastic ones.

Poly-pots do not take up much space when stored flat, and tapered plastic containers usually stack conveniently inside one another.

Should I use nursery soil to fill the pots?

Occasionally yes, if it is **really** suitable for the purpose; ***but***
Generally no, because a **potting mix** will give much better results.

But young trees planted in the nursery would grow all right!

Yes they might, but with the restricted rooting conditions in a container you would generally get poor planting stock and many losses, because the young trees:

(A) could easily get too little water, or too much (C 13, C 43);

(B) might have their roots deprived of good aeration (C 11);

(C) may not receive enough nutrients to make reasonable growth (C 14, C 33); ***and so***

(D) are unlikely to produce a good, well-branched root system and sturdy top (C 34).

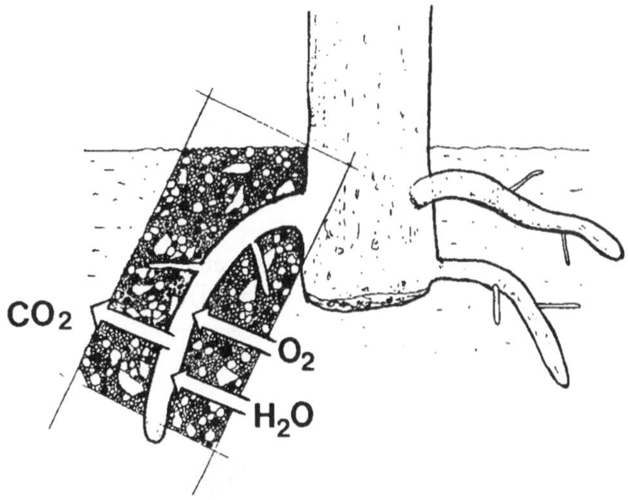

O_2 = oxygen
CO_2 = carbon dioxide
H_2O = water

What is a good potting mix like?

Good growth of the young trees generally occurs in a mixture that:

(1) has some coarse as well as **smaller soil particles** in it (D 12 in Manual 4), to provide adequate but not excessive **air spaces** (with an *aeration porosity* of 25-35% - see Landis et al., 1990 in C 61-B), so that:

(a) *oxygen* from the air can reach the roots easily;

(b) excess water drains away freely, not leaving the soil waterlogged (C 11);

(c) the roots can grow and branch without restriction; ***but***

(d) the root ball holds together after the pot has been removed;

(2) contains enough **organic matter** and **nutrients** to:

(a) retain sufficient water within the soil between successive waterings;

(b) release nutrients gradually (C 33-34), partly through the activities of *decomposers* in the soil (D 13 in Manual 4); ***and***

(3) is neither **alkaline** nor **strongly acid** (the desirable pH is usually between 4.5 and 6.5).

Most nursery soils (C 23) cannot perform all these important functions, though a few can form part of a suitable potting mix.

Isn't it hard to get a soil with all those features?

No, it is usually quite easy, using locally available materials. *Mixing these together in suitable proportions* is the way to obtain a good potting soil.

What components are useful?

(1) Coarse sand and fine gravel from a river or beach, washed to remove salt or other impurities;
(2) Loamy topsoil, collected in closed woodland rather than from open ground;
(3) Sawdust or bark chippings that have been allowed to weather for about a year, so that any toxic substances have been broken down (C 33);
(4) 'Black' soil from waste tips, sieved to remove glass and other undesirable items;
(5) Coconut or sugar-cane fibre that has partly rotted;
(6) Rice hulls: ***and***
(7) Composts of various kinds of waste vegetable matter and/or animal droppings (C 33).

Note: potting soils sometimes need the addition of small quantities of fertiliser if certain nutrients are lacking (C 14, C 33), or inoculation with micro-organisms (C 30-32) if particular tree species are to thrive.

How do I decide what potting mix to use?

You could:

(A) start with potting mixtures that have already been found satisfactory; ***and***
(B) do some simple pot experiments (C 7, C 15) to compare them with others that might produce better planting stock, such as:
(1) 20% coarse sand, 40% loamy topsoil, 30% weathered sawdust, 10% compost;
(2) 25% coarse sand, 25% loamy topsoil, 25% rotted coconut fibre, 25% black soil.

The proportion of coarse sand needs to be reduced when the loam is sandy, and increased with clayey loams.

Are there some well-tried examples?

The booklet by Josiah and Jones (1992) listed in sheet C 61-B includes details for making potting mixes from waste sugar-cane (developed in Haiti), coconut husk (Thailand) and pine bark (South Africa).

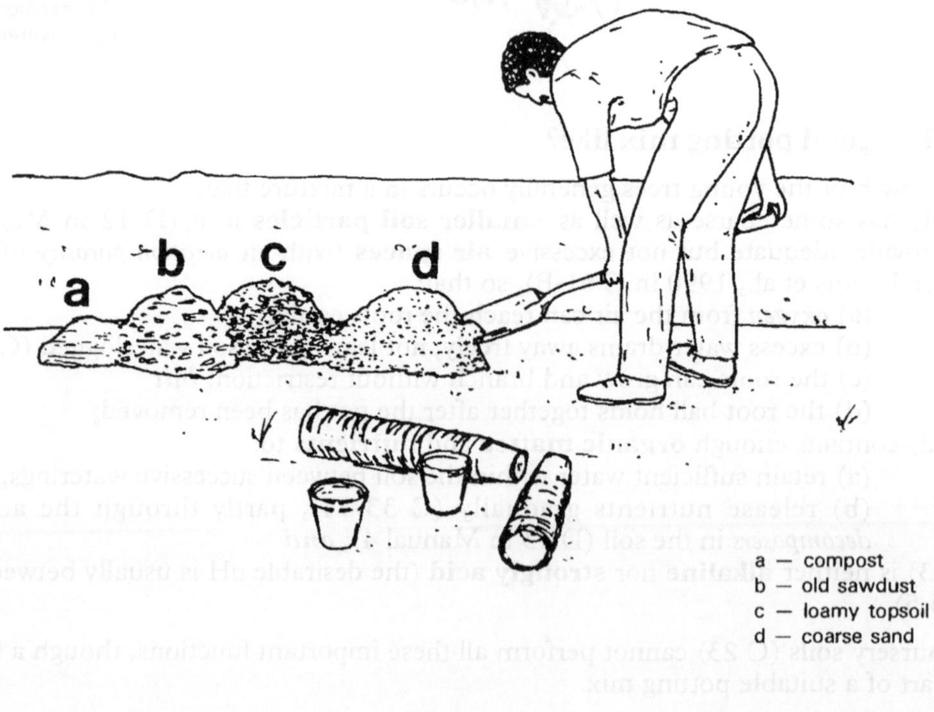

How do I mix the components?

(1) Wash coarse sand and fine gravel beforehand;
(2) Sieve items (2) - (7) above, to remove any large lumps or stones, and sharp items that might cut people's fingers and perhaps transmit disease;
(3) Add the required amounts of each component (for example, to make 10 bucketfuls of potting compost (B 1) above, add two bucketfuls of coarse sand, four of loamy topsoil, three of weathered sawdust, and one of compost); ***and***
(4) Mix everything together thoroughly.

Preparing the potting mix is best done with a shovel on a flat surface such as concrete, hard soil or an old sheet of plywood. Small quantities can be mixed with a trowel on a table or workbench, while a cement-mixer can be handy for large quantities, especially if uniformity is needed for an experiment (C 7).

And what about water?

The mix should not be ***too wet,*** and so:
(a) wash the sand and allow it time to drain before using it; ***and***
(b) cover components or keep them inside a shed to protect them from rain; ***but***
Nor should it be ***too dry,*** and so:
(1) water some of the stored components occasionally during dry weather; ***and***
(2) add water gradually to the potting mix until it is **moist** but not wet.

Is a potting mix also suitable for sowing seeds?

Yes, if it is finely-sieved and not too rich in nutrients (C 14, C 33), it can be put in seed trays for smaller sized seeds, or into pots for larger seeds.

What should not be put into a soil mix?

(a) Stones or large pieces of gravel;
(b) A lot of clay;
(c) Salt;
(d) Fresh manure or sawdust;
(e) Broken glass or thorns;
(f) Numerous weed seeds;
(g) Known pests or harmful micro-organisms.

How about potting up the young trees?

This is covered in sheet C 42.

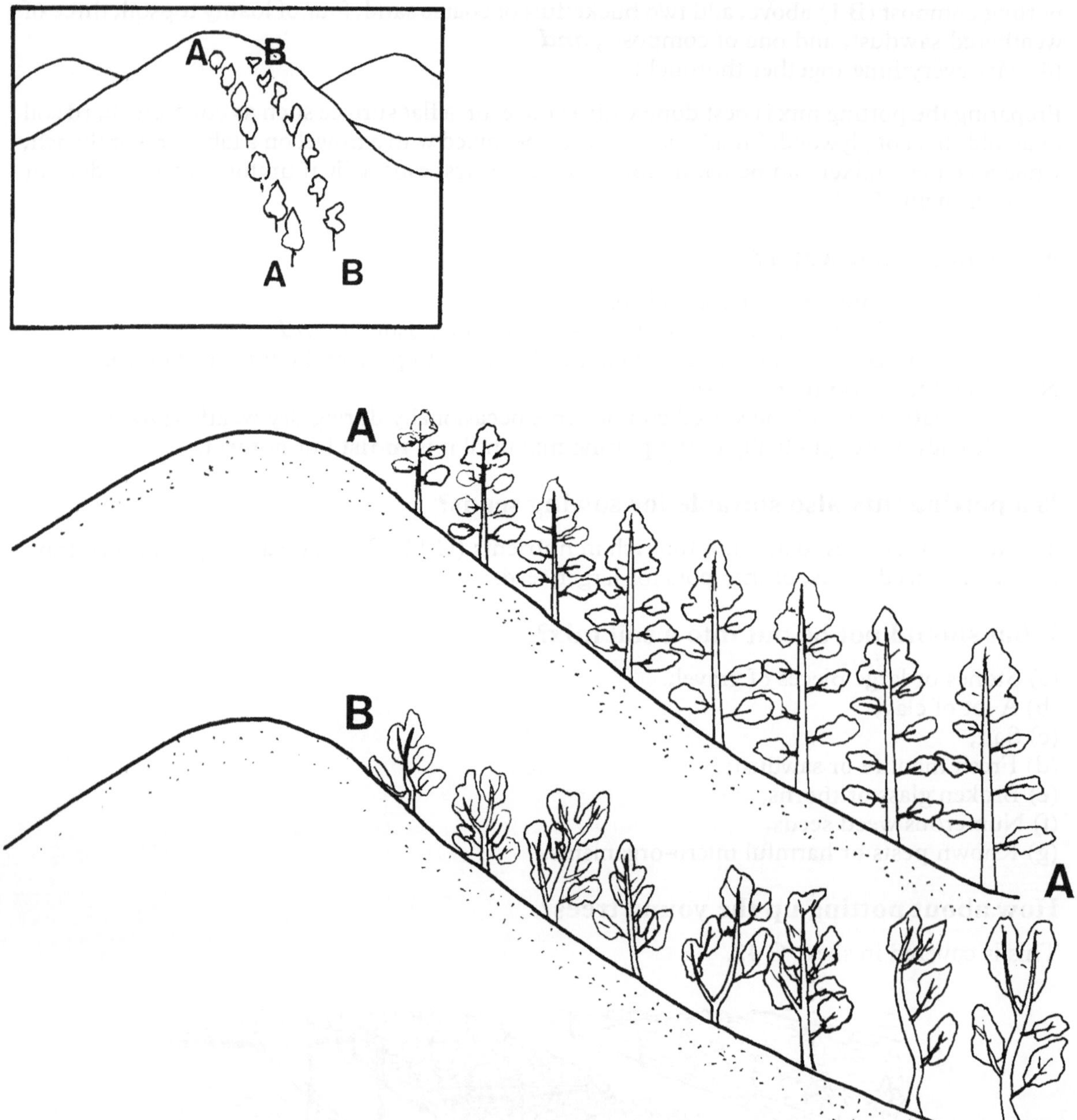

Clonal research.

Introduction — C 7
- *growing trees for research*

Is research on trees really required?

Yes it is (C 4). Relatively little work has been done in the tropics on:

(1) the many factors that influence tree growth (C 10-15) and survival (D 10-16 in Manual 4), and how they interact with one another (C 69-K);

(2) the ways in which the different aspects of tree growth are internally controlled and co-ordinated (C 14); ***or on***

(3) propagating and planting the great majority of tree species.

Which kinds of research are particularly needed, then?

Many different approaches, especially formal and informal experiments (D 6 in Manual 4). In Universities, research stations and international projects, it is important that some basic studies of (1) and (2) above are included. Since many aspects of tree biology are little known for many species, these could include important subjects like:

(A) possibilities of vegetative propagation (C 5; and Manual 1);

(B) times of flowering and fruiting (Manual 2); ***and***

(C) seed physiology in relation to seed handling.

What might practical tree growers and planters concentrate on?

Investigating for instance:

(a) nursery problems and new techniques in **potted plant experiments** (C 15);

(b) establishment problems and new methods in **planting experiments** (Manual 5);

(c) the performance of new species and genetic selections in **field trials** (D 29 in Manual 4).

Are there any hints on useful treatments to apply?

(A) **Potted plant experiments** (C 15) might for instance examine, for a species that is slow or difficult to grow, the effects on root and shoot development of various:

(1) potting soils or containers (C 6);

(2) levels of shading, or the rate at which shade is reduced (C 41, C 47); ***or***

(3) frequencies and amounts of watering (C 43), or the spacing of containers (C 42).

(B) **Planting experiments** need uniform batches of good young trees, and could investigate such subjects as:

(1) the best sizes for planting stock (C 34);

(2) tree growth and establishment on contrasting sites (D 20-29 in Manual 4); ***or***

(3) planting techniques or timing (Manual 5).

(C) **Field trials** (both formal and informal) could explore matters like:

(1) the growth and form of various promising species;

(2) which provenances, seed-lots or clones perform best (C 5); ***and***

(3) spacing, thinning and yields.

What's different about growing trees for research?

One needs to follow the best available methods for growing ordinary planting stock, but to take extra care (C 48) in order to produce batches containing plenty of good, uniform trees that have not been subject to marked climatic or other stress (C 41).

How can I grow uniform sets of young trees?

(a) By selecting from a restricted genetic range;

(b) By propagating and growing all the batch in very similar conditions; and

(c) By producing more than enough, and then excluding the biggest and smallest trees, as well as poorly growing or unhealthy plants.

In what ways could I restrict the genetic range?

(1) By collecting **seed** from one or a few closely-related mother trees (C 5); ***or***
(2) By using one or a few **clones** (A 10-11 in Manual 1).
If neither is available, use a single seed-lot (or if necessary 2-3 that have been well mixed).

But isn't it a mistake to have too little genetic variation?

Yes** for planting stock,* though neither should there be too much (C 5); ***but
***No** for experimental trees,* except when studying the range of *inherited* differences.

Why should I use a narrower genetic base for most experiments?

Because it reduces the overall variation in the experiment, making it more likely that *significant differences* will show up between treatments (C 55, C 69).

But will the results apply to the whole species?

Strictly speaking, the results of any experiment apply to the time it was done, the particular plants used and the conditions they experienced. This is one of the reasons why a series of simple, straightforward experiments can be more informative than a single complex trial. Try and build up reliable information by:

(a) starting with strongly limited genetic variability;
(b) concentrating on finding out whether the chosen treatment has any effects on the growth of the young trees; ***and***
(c) leaving the question of how representative the results are until a later stage.

In due course, one might then be able to use several clones, derived from provenances (C 5) across the whole range of the natural distribution of the species.

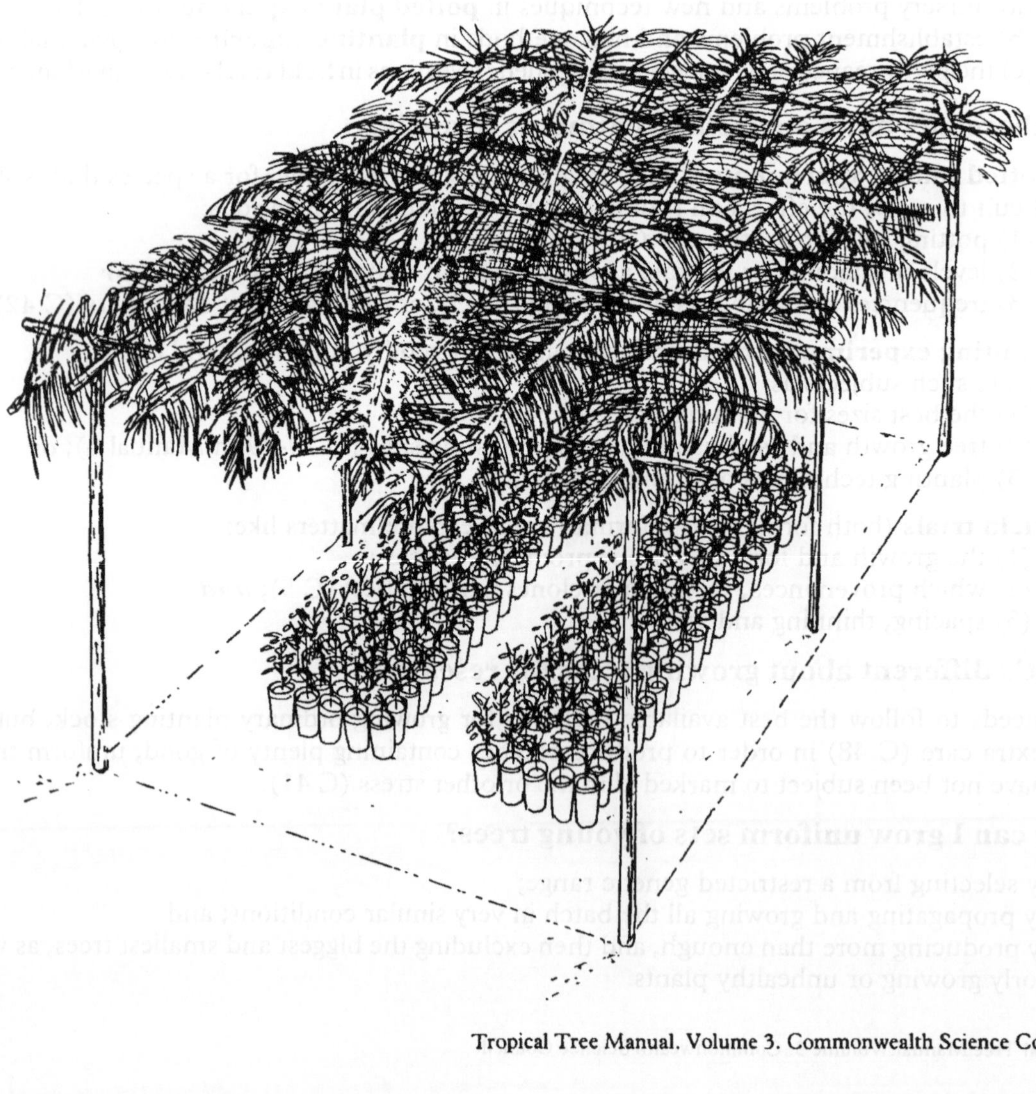

How should I try and maintain similar conditions for growth?

By raising the experimental trees from either:

(1) seeds sown on the same day in similar germination conditions (Manual 2); ***or***

(2) cuttings that were taken and rooted at the same time (Manual 1);

Growing them in:

(3) containers of the same type, size, shape and colour (C 6);

(4) a good potting soil, all of which has been thoroughly mixed together; ***and***

Keeping them under:

(5) light and temperature conditions that are as uniform as possible, such as a large area of uniform shade, a special shade house (C 48) or full sun;

(6) particularly careful checking (C 40) and watering (C 43) regimes; ***and***

(7) protection from wind, fire, grazing animals and vandals (C 46).

Whenever possible, set up the experiment in one day, and have the same person doing a particular task on all the trees.

Should I add extra nutrients to the potting mix?

Not unless it is needed, for example:

(a) to obtain reasonable growth rates in poor soil (C 33);

(b) to have the trees ready for the date of planting out (C 34, C 47); or

(c) if the young trees are going to be used for experiments in containers (C 15), where they could be given equal amounts of a dilute liquid fertiliser at intervals.

What else is needed?

(1) Extra planning beforehand (C 50, C 54, C 62-F);
(2) More regular weeding (C 44);
(3) Fuller records (C 55, C 64-67).

Why exclude the biggest trees from the experiment?

Because it helps to have experimental trees of more or less the same size.
If the batch is not large enough for such 'standardising', you could 'grade' the trees into size classes (A 45 in Manual 1).

And how will I judge the results of the experiments?

For potted plant experiments, see sheets C 55 and C 67-69;
For planting experiments and field trials, see Manual 5.

Avoiding bias when using research plants twice.

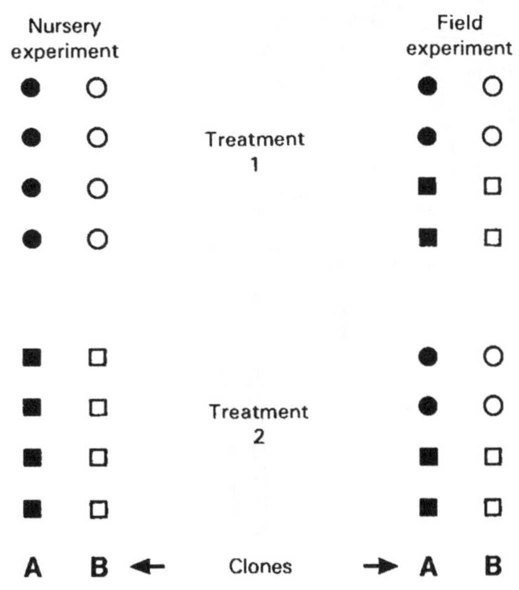

Could I use plants from a potted plant experiment for a field trial?

Yes, if they are in reasonable condition, and:

(A) other suitable young trees are in short supply; ***or***

(B) they are clonal plants that are particularly appropriate.

However, make sure that the labels are retained (C 54), as the new experiment will have to be laid out 'across' the old one, to avoid bias and improve precision.

How would that be done?

The surviving healthy plants which formed the controls in the potted plant experiment would need to be equally divided between each treatment of the field trial. The same would then be done with the trees from each of the other potted plant treatments.

If sets of plants of different genetic origin were used in the potted plant experiment, a similar procedure would need to be followed so that they were evenly spread amongst the treatments in the field trial.

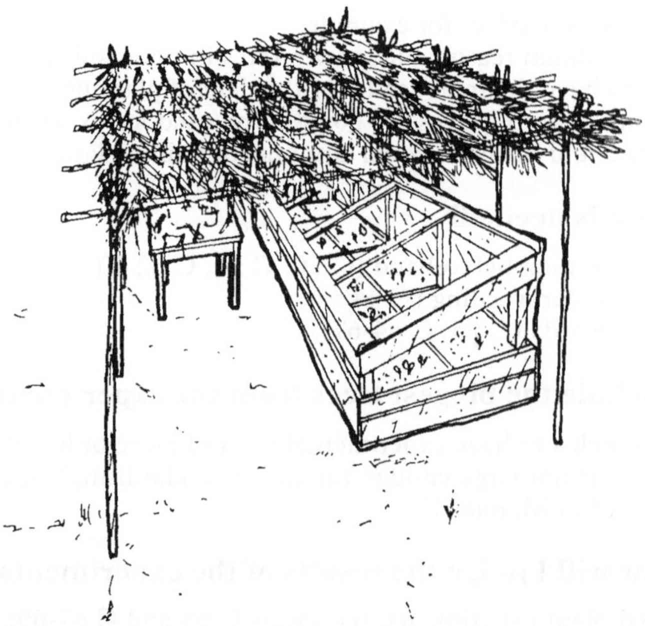

GENERAL PRINCIPLES OF TREE GROWTH

General principles of tree growth - *introduction: how trees grow* C 10

How do trees grow?

By using the energy of sunlight to make organic matter out of simpler substances.

Does the sunlight just warm them up?

No, although it does do that. Like all green plants, trees contain a pigment called *chlorophyll* which absorbs some of the sunlight, allowing part of its energy to be turned into a chemical form.

During the second stage of this process of *photosynthesis*, the chemical energy is used to build in *carbon dioxide* absorbed from the air to form sugars (D 10 in Manual 4).

What happens to these sugars?

Of the large amount produced in a tree, some sugars are:

(a) **broken down again** in the process of *respiration*, providing useful chemical energy throughout the tree (and releasing carbon dioxide again);

(b) **built up** into many other chemical substances that allow it to live, grow and form roots, trunks, branches, leaves, flowers, fruits and seeds; ***and***

(c) **stored**, often as starch.

Does the tree keep on storing more and more?

Yes it may do so, if there is enough light. However, quite a lot of its sugars, starch and structural parts are:

(1) eaten by animals, which cannot produce their own food; ***or***

(2) shed from the tree as dead leaves and other parts, and ripe fruits. These form *litter*, which is then broken down by *decomposers* (D 13 in Manual 4).

Tropical trees are often the main **producers** in the **food chains,** on which all the **consumers** (most other living organisms) depend (D 10).

But what has all this got to do with tree nurseries?

Growing good trees successfully is determined by the general principles which control tree growth. Practical experience is important too, but it is most effective when combined with a grasp of the basic elements of tree biology.

How does a tree actually get bigger?

It does so because some of its *cells:*

(A) ***divide*** into two, making more cells, some of which may themselves divide; ***and***

(B) ***increase in size*** considerably, so that they become much bigger.

What are cells?

They are the living units of which the tree is made, which often:

(a) expand mainly ***in one direction,*** causing that part of the tree to grow in a specific way;

(b) ***become specialised,*** so that groups of them look different and have a variety of functions; ***and***

(c) can ***continue*** to function even ***when dead.***

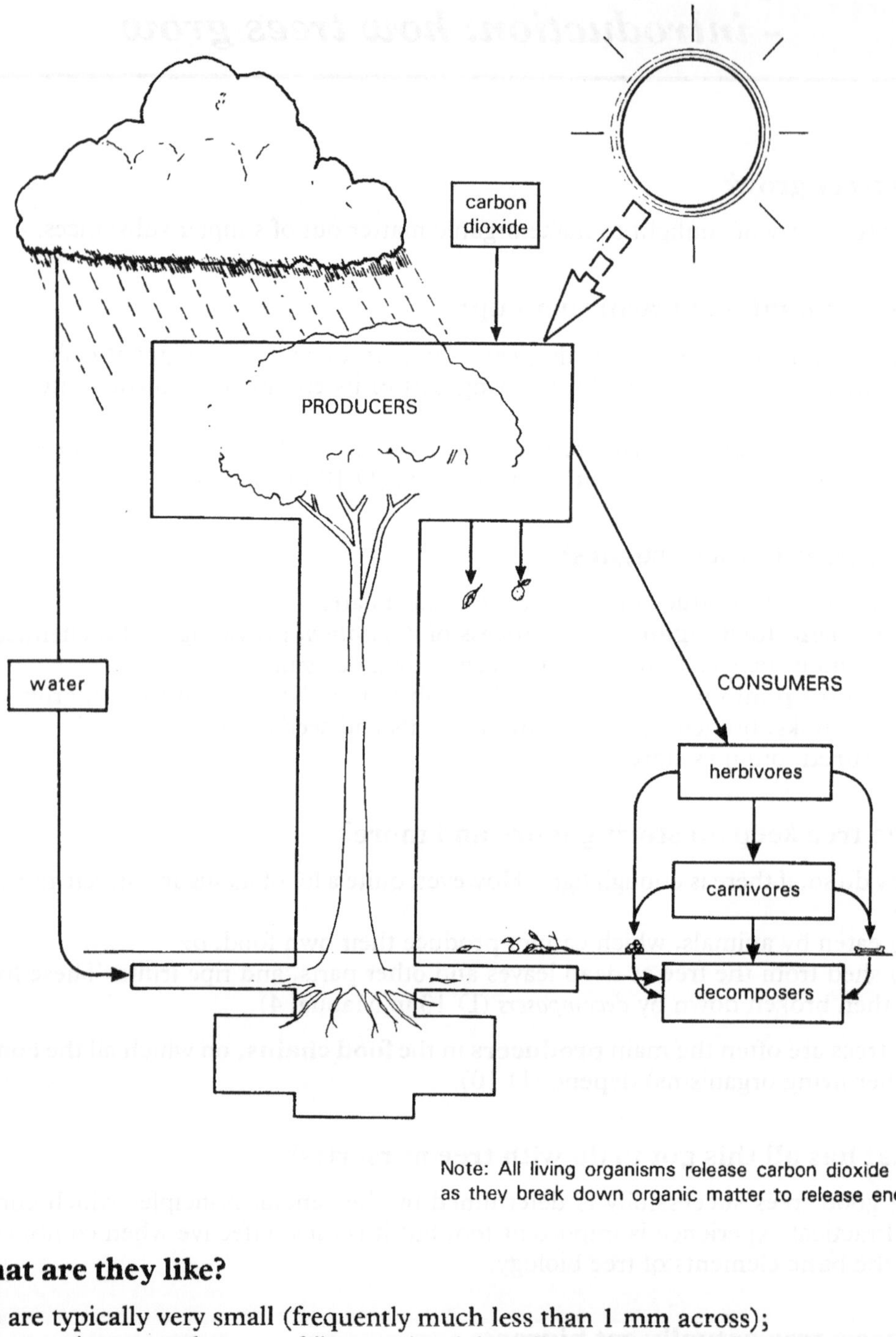

Note: All living organisms release carbon dioxide to the air as they break down organic matter to release energy.

And what are they like?

(1) Cells are typically very small (frequently much less than 1 mm across);
(2) They contain a complete set of 'instructions' for ***all*** the things which can be done by that individual tree (C 17); ***but***
(3) They carry out only some of these functions.

For example, the cells in the root possess all the instructions for making fruits, but these remain unused.

Why is that important?

Because it means that:
- (a) damaged parts of the tree can often be repaired;
- (b) stem cuttings can produce new roots (Manual 1); ***and***
- (c) cut stumps can produce coppice shoots.

In theory, any living cell could be turned into a whole tree.

How many cells are there in a big tree?

A single leaf can contain several hundred, and a large tree many millions.

Well, why not turn a good tree into millions of young ones?

Unfortunately, there are a lot of barriers to such *micropropagation* (A 5 in Manual 1). For example:

(A) many of the cells in a large tree are dead, while others have lost the ability to divide;

(B) most of the living cells could not be separated from one another without killing them;

(C) cells taken higher than a metre or two above the ground are probably no longer *juvenile* (C 5; and A 21 in Manual 1), and therefore might not produce good trees;

(D) culturing individual cells and persuading them to produce a lot of new *plantlets* is not always easy; ***and***

(E) there can be genetic changes and losses of parts of the instructions during this kind of micropropagation.

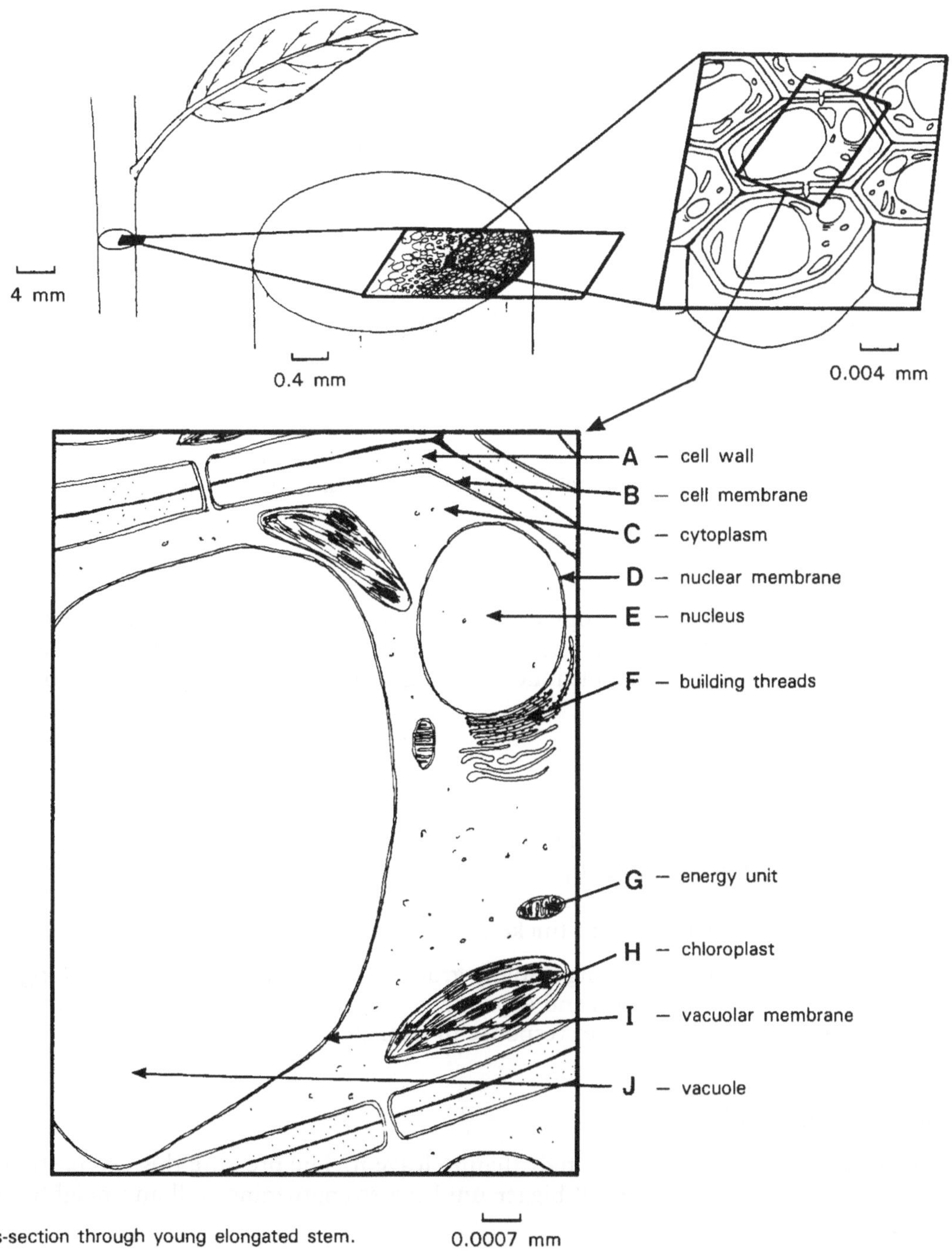

Cross-section through young elongated stem.

Does that mean that micropropagation can't help?

No it doesn't, because:

(1) culturing **detached shoot tips** appears to have better prospects than starting with individual cells, and problems (D) and (E) appear to be less serious; ***and***

(2) the potential for multiplying the numbers of young trees for planting could still be large in tree species that produce a lot of shoot tips, and in those which can be persuaded to form many shoot tips or *somatic embryos* in culture.

However, for most tree species the best prospects are to root leafy, juvenile cuttings in a shaded poly-propagator (Manual 1), and grow them on in a tree nursery, because this approach:

(a) can easily be fitted into existing nursery practice;

(b) does not need expensive facilities or depend on an assured electricity supply;

(c) makes it easier to use superior selections to produce clonal planting stock (C 5).

What are cells like inside?

The living part usually includes:

(1) **the cell membrane,** a very thin layer on the outside which governs the movement of substances into and out of the cell (C 14);

(2) **a nucleus,** which contains the genetic instructions, and controls the activities of the cell;

(3) **cytoplasm,** which carries out the first part of respiration, and contains various structures including:

(a) **energy units,** which perform the second part of respiration, releasing useful chemical energy;

(b) **building threads,** where the essential *proteins* are made.

Cells may also contain:

(5) **a vacuole,** an inner sac of water and dissolved nutrients that often takes up most of the space; ***and***

(6) **chloroplasts,** which are the green, chlorophyll-containing units that carry out photosynthesis.

Outside is the **cell wall,** which limits the tendency of the living part to go on getting bigger, and if strengthened may give support to the tree even after the cell has died.

How do cells divide?

The instructions in the nucleus are copied **exactly,** the two versions separate, and new cell membranes and walls are formed between them. Most cell divisions result in two identical daughter cells, one of which often continues dividing while the other expands and becomes specialised.

Can any cell divide?

Any living cell can theoretically do so, but in practice most cell division happens:

(a) in the growing points at the tips of roots and shoots, making them get longer;

(b) between the wood and the inner bark in main stems and branches (C 12) and in some roots (C 11), making them thicker.

Cell division also occurs in other young, rapidly growing parts of the tree, such as elongating fine roots, extending stem internodes and expanding young fruits. It also takes place when new growing points are formed, and in response to damage.

How do cells enlarge?

Water tends to enter the cell more strongly than to leave it. The pressure this exerts stretches the new thin cell wall, making the cell bigger until the strengthening wall and neighbouring cells restrain the enlargement.

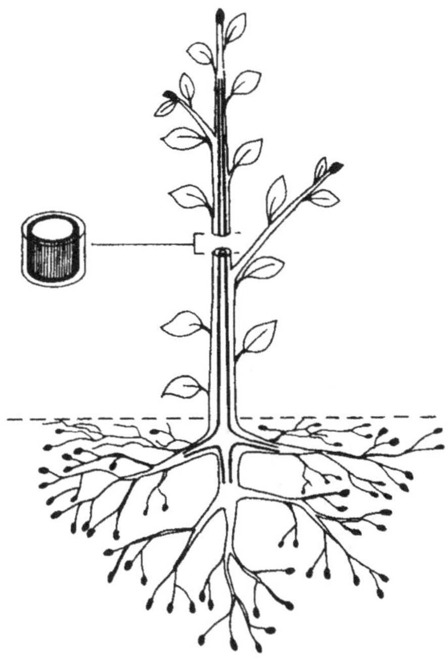

Where most cell division happens. Stylised young tree, showing many root tips, three shoot tips, and the layer between wood and inner bark.

What kinds of specialised cells are formed?

Some common examples are:

(1) **sugar-making cells,** which contain a lot of chloroplasts, and are found especially in leaves and near the surface of young stems and unripe fruits;

(2) **sugar-conducting cells,** which are long tubes that remain alive for some time. They connect up different parts of the tree, and are common in the inner bark;

(3) **water-conducting cells,** which quickly die and lose their contents, often becoming like long narrow pipes. They also connect different parts, and are present in large numbers in the sapwood of roots, trunk and branches;

(4) **surface cells,** that:

(a) *in the fine roots* often absorb water and nutrients from the soil (C 16); ***and***

(b) *in above-ground parts* restrict drying up of the tree, but permit carbon dioxide and oxygen to enter (C 12);

(5) **storage cells,** which can be full of starch grains, and in seeds may also contain stored proteins and fats; ***and***

(6) **strengthening cells,** whose walls are often thick and strong.

Don't things go wrong sometimes?

Yes, occasionally a cell is formed that has incorrect instructions. However, in most cases there are specific structures in the cytoplasm which destroy such faulty cells. The contents are broken down and other cells take over.

What affects growth and division of cells?

(A) Where they are situated within the tree;
(B) Its hormonal control systems;
(C) The availability of water, oxygen, sugar and nutrients;
(D) Temperature; ***and often***
(E) Light.

Some of these effects will be mentioned in sheets C 11-15 and C 34.

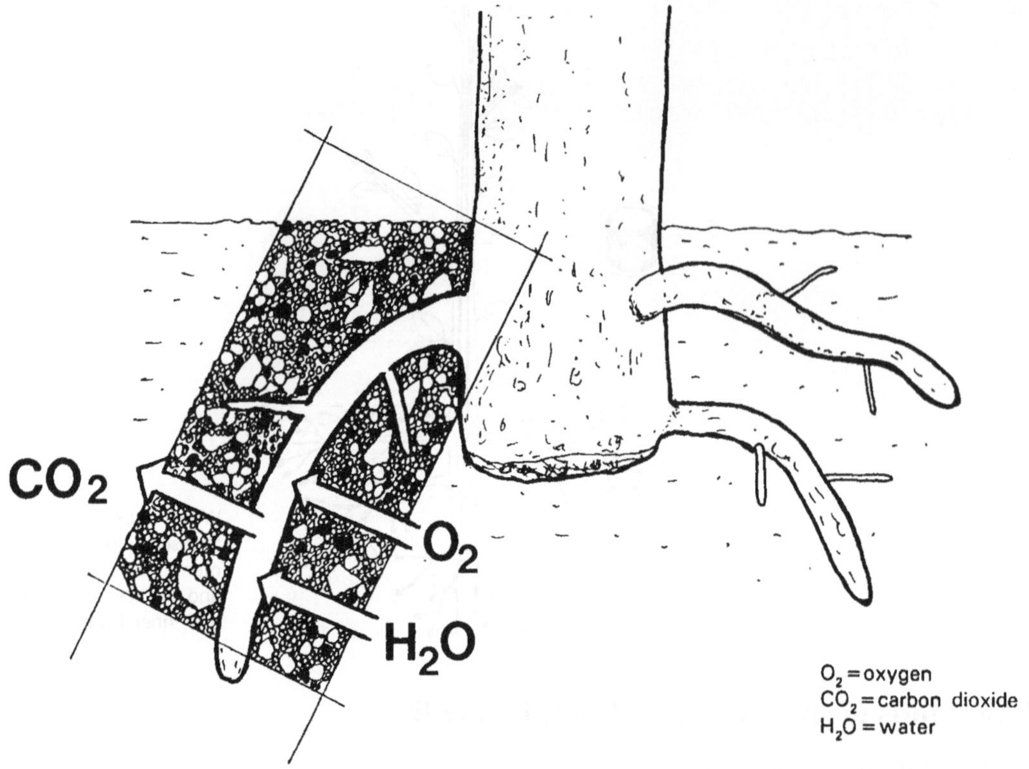
CO2
O2
H2O
O2 = oxygen
CO2 = carbon dioxide
H2O = water

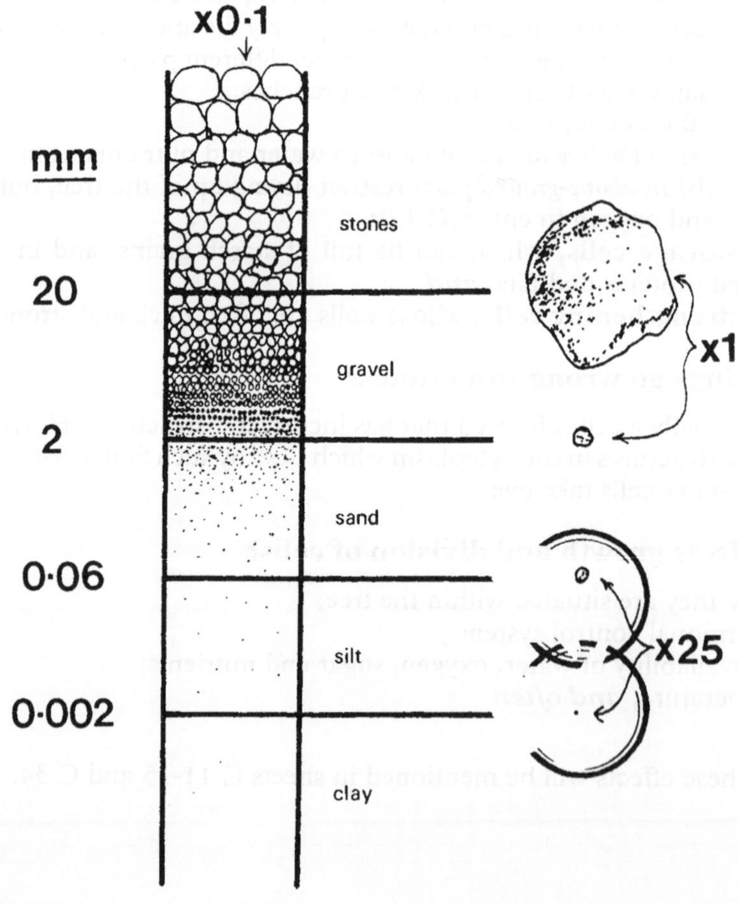
x0·1
mm
stones
20
gravel
x1
2
sand
0·06
silt
x25
0·002
clay

General principles of tree growth - *expanding root systems* C 11

What kinds of roots do trees have?

In a young nursery tree, the root system generally consists of:

(1) **main roots,** including the taproot and others arising from the root collar or the base of the stem;

(2) **branch roots** formed at the sides of the main roots, which in turn branch frequently.

In a larger, well-established tree, there are often:

(a) thickened **structural roots** in the soil, going downwards but also spreading sideways;

(b) fine **absorbing roots,** growing and branching repeatedly in the top 10 cm;

(c) many **mycorrhizas,** close associations between fine roots and fungi (C 30-31).

In some tropical trees part of the root systems are above-ground, for example as *stilt roots* or as *'breathing' roots,* found especially when the soil is waterlogged.

How does a root grow?

By some of its cells dividing, becoming larger and more specialised (C 10). Within one actively growing root, there are usually zones where cells:

(1) **are continually being formed,** in the root tip, just beneath the root cap;

(2) **elongate greatly,** while some of them continue to divide. This zone, about 2-20 mm behind the root tip, pushes it forward through the soil;

(3) **absorb water and nutrients,** in the parts that are no longer extending, often in association with fungal threads running through the soil, and sometimes root hairs. Branch root tips are often formed in this part; ***and***

(4) **conduct water and dissolved substances,** in the parts that are becoming thicker and no longer absorb from the soil.

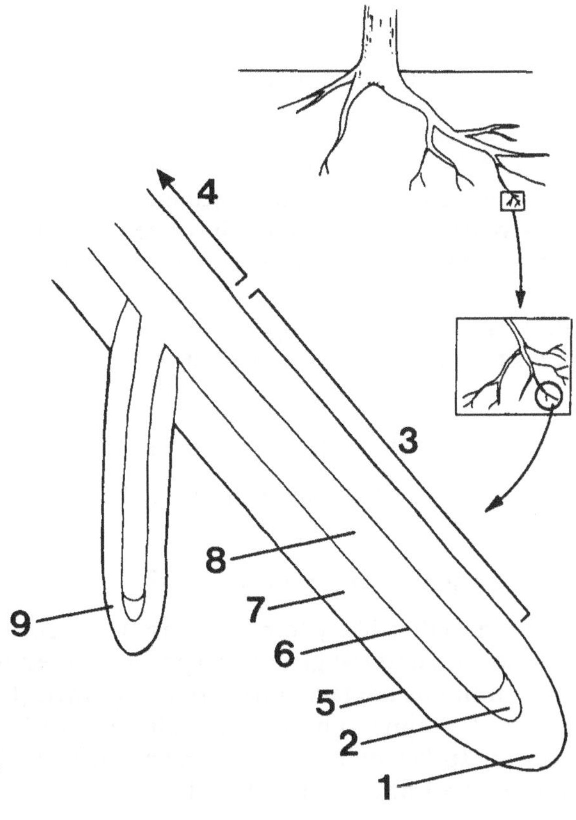

Section through a young root.

1 — root cap. 2 — cells dividing.
3 — cells dividing, elongating and becoming specialised.
4 — zone no longer elongating.
5 — surface cells. 6 — boundary cells.
7 — storage cells. 8 — conducting cells.
9 — branch root.

Do all roots start off the same?

In some tree species, young root tips are of two distinct types:

(A) **thick roots** (1-2 mm in diameter), which may become structural roots and continue to grow in diameter; ***and***

(B) **thin roots** (less than 1 mm in diameter), which may function as short-lived absorbing roots that remain thin but branch frequently.

In seedlings (Manual 2), the first root is generally thick, making branch roots that may be thick or thin;
In cuttings that have been treated with *auxin* (A 40 in Manual 1), both types may be formed, but without added auxin only thin roots may sometimes be produced.
In palms, the roots are usually very thick when formed, and remain the same diameter, without branching.

How do the roots of other trees get thicker?

By producing a cylindrical layer of cells that then divide continually, forming water-conducting and strengthening cells to the inside and sugar-conducting cells to the outside (C 10). Because this layer of dividing cells lies deep inside the tissues, the root can actually get thinner as the outer parts are lost, before thickening later on. Like stems (C 12), thickened roots also produce bark to the outside.

And how do branch roots form?

A small group of cells divide and form into a root tip, which then grows in the normal way, bursting out into the soil through the outer tissues of the original root. Older, thickened roots can also produce branch roots, especially after disturbance or injury.

When cuttings form roots (Manual 1), a similar process happens near the base of the stem, which is stimulated by adding small amounts of auxins.

Can I increase root growth by adding auxins?

No, applying these hormones will make root *elongation* slow down or stop altogether.

Do roots normally grow all the time?

Growth in length of roots is often continuous in the tropics, though with periods of faster and slower growth. Similarly, there are often peaks in the formation of new thin roots. However, roots might stop growing completely if subjected to:

(A) **serious wilting** - when water has been lost from the shoot faster than the root system can absorb it (C 13). If the shoot is beginning to wilt, the root cells will also be short of water, and may also produce different hormones (C 14); ***or***

(B) **chilling injury** - if soil temperatures fall to around 10°C, the root cells of lowland tropical trees may be seriously damaged (C 41; and D 11 in Manual 4).

Is growth in thickness of roots continuous?

It generally is, unless the trees become leafless (C 12).

What influences how fast the roots grow?

In the nursery, root growth is likely to be most rapid when:

(1) both the species and the genetic origins are suited to the locality;
(2) the soil conditions are favourable to root growth (C 6, C 23);
(3) the tree has had time to become established in a bed or container;
(4) it is receiving sufficient light (C 41) and nutrients (C 33-34), is being carefully watered (C 43) and well looked after (C 40, C 48).

What other things slow down root growth?

Root growth will tend to decrease or even stop if, for example:

(A) the trees are kept in such dense shade that sugar production in the shoots barely exceeds what is used up or lost by the whole tree (C 10);

(B) the tree is subjected to a lot of *stress*, for instance from lack of or too much water, low or very high temperatures, strong winds or toxic chemicals (C 41);

(C) roots are badly damaged during potting up (C 42); ***or***

(D) the roots in a container become *pot-bound,* just going round and round outside a poor or exhausted potting soil (C 6).

Why is too much water bad for root growth?

Because the water drives out the air from the spaces between the soil particles, so that the root cells run short of oxygen. Such conditions can also favour micro-organisms that cause diseases (C 45). Waterlogging can happen if:

(1) the drainage holes in pots are too small (C 6), or get blocked by soil or roots;

(2) the potting soil contains a lot of clay;

(3) it becomes compacted inside the container; ***or***

(4) a newly-potted young tree is subjected to prolonged rainfall or is over-watered.

Tree species from mangrove woodland, freshwater swamps, and seasonally flooded forest and savanna generally thrive in soils liable to flooding. Some of them have:

(a) aerial 'breathing' roots reaching above the water level;

(b) air-conducting passages that allow oxygen to reach the root system; ***or***

(c) other special adaptations for coping with lack of oxygen.

But isn't the purpose of roots to supply water?

Well, it is one of their four vital functions, which are:

(A) taking up water;

(B) absorbing nutrients;

(C) manufacturing substances; ***and***

(D) anchoring the tree into the ground.

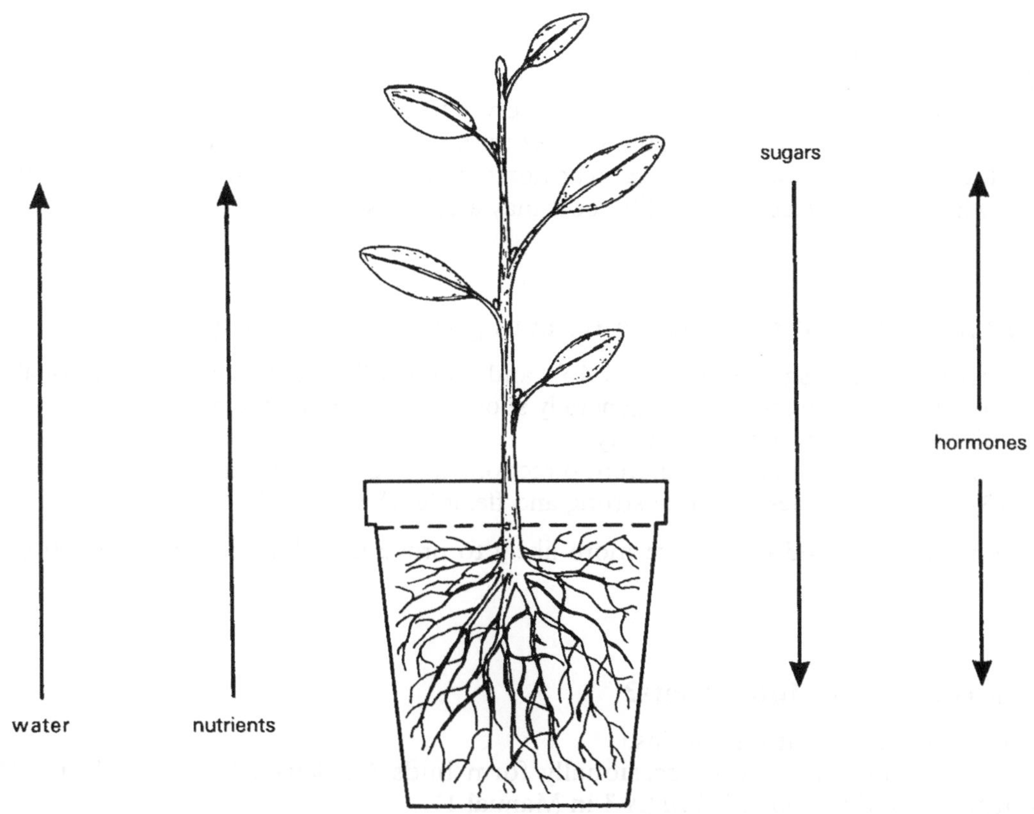

How is the water actually taken up by the tree?

When a fine, absorbing root is in close contact with small, moist soil particles, it is easy for water to pass through the thin walls of its surface cells. Because of the dissolved substances they contain, water is then taken up strongly through their cell membranes (C 10). Most of it moves from cell to cell into the water-conducting 'pipes' of the root, and then into the shoot system.

Water may also pass into the roots from the extensive network of fungal threads of mycorrhizas (C 30-31).

Are nutrients taken up the same way?

If the nutrient is dissolved in water, it can enter the cell walls easily. However, the next stages are different (C 14), because:

(1) the cell membrane is **selective,** not taking up nutrients in proportion to their concentration in the soil solution; ***and***

(2) uptake of nutrients requires the use of chemical **energy** (C 10) to transfer them across the cell membrane.

Once absorbed, the nutrients may be:

(a) used for growth in the root;

(b) stored; ***or***

(c) transported to the shoot either in the water- or sugar-conducting cells.

What sort of substances are manufactured in the roots?

Using sugars arriving from the leaves, the root system usually:

(A) produces the many chemical substances for the new root cells;

(B) stores starch in some of them;

(C) makes specific hormones that help to correlate the growth in the shoots with that in the roots (C 14).

How about anchorage?

Starting in the nursery, the root system also serves to fix the tree into the ground, keeping it upright and stopping it from being blown over (and in flooded sites from being washed away). When the tree is older, a set of large near-horizontal roots with *sinkers* may anchor the tree firmly, while in other tree species stilt-roots may assist in stability.

Aren't nursery root systems too weak to give much stability?

Young trees do need shelter (C 25, C 46), and could still be damaged by occasional fierce storms (C 3). However, their roots generally soon provide anchorage because:

(1) root numbers increase rapidly;

(2) roots branch in many different directions through the soil; ***and***

(3) the thicker ones are quite strong and flexible when pulled.

Poor, diseased or spindly trees are more likely to prove unstable than good planting stock (C 4).

Do roots ever turn into shoots?

Root tips do not turn into shoot tips.
Certain cells on older roots occasionally form buds (suckers), for example in *Milicia (Chlorophora)* and *Cordia alliodora* (A 3 in Manual 1).

Remove section of root

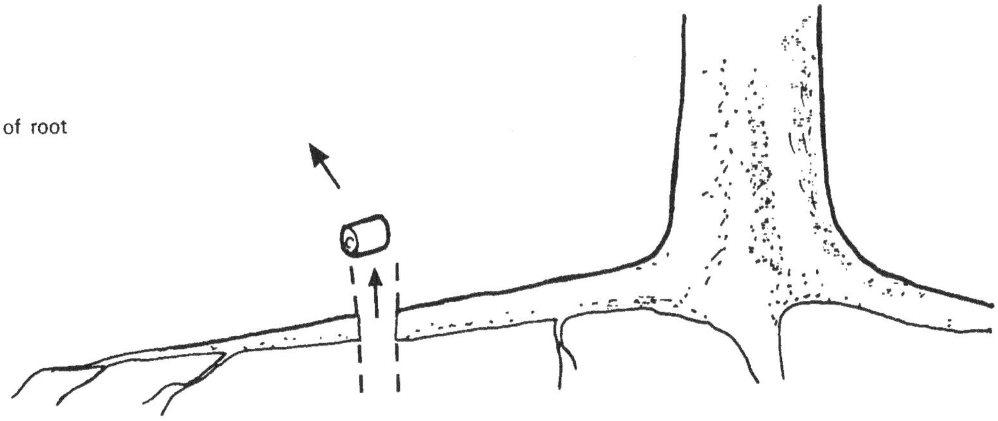

Dig up sucker + root

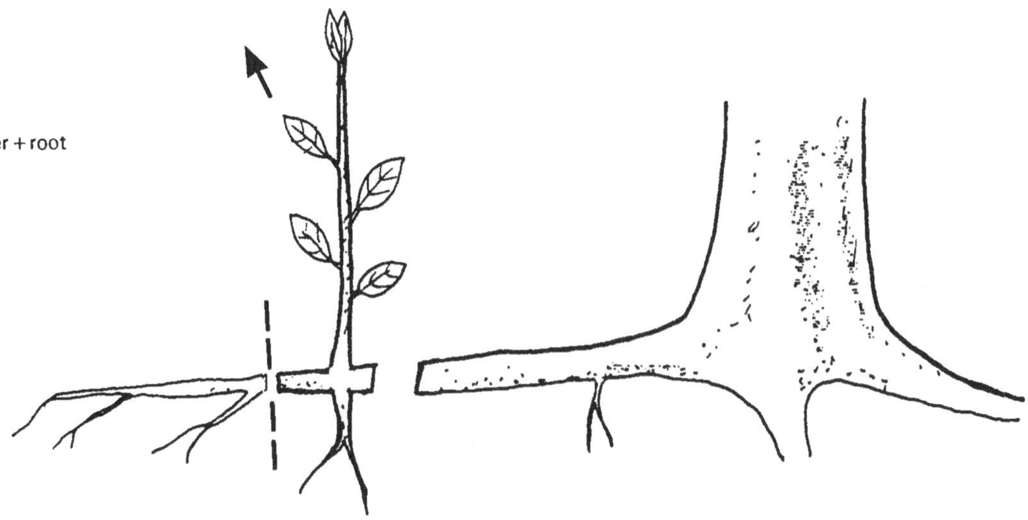

Can root systems exist by themselves?

Only in the laboratory, and when provided with a sugar supply. Roots usually contain few or no sugar-making cells (C 10), and so are just as dependent upon photosynthesis in green shoots as are humans, other animals and decomposers (D 10 in Manual 4).

Does that mean that the shoots are more important?

No, because:

(a) they in turn are dependent upon the roots, both for supplies of water and nutrients and for stability; ***and***

(b) a good nursery root system is more crucial to tree survival after planting than the sort of shoot (C 34).

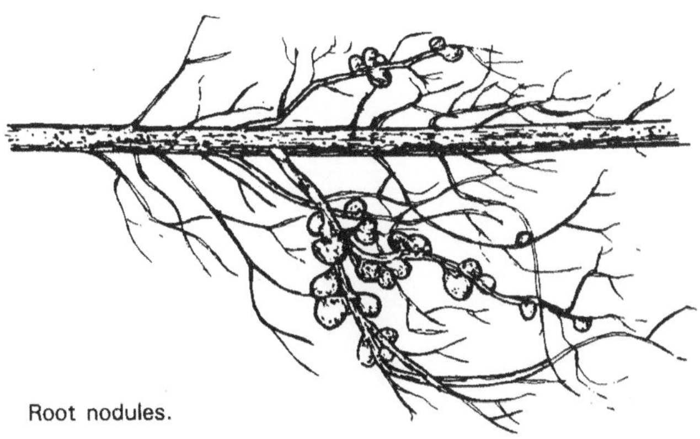

Root nodules.

General principles of tree growth C 12
- growing stems, leaves and branches

How do shoots differ from roots?

In the way they are organised, how they grow and what they do.

Do they have the same kinds of cells?

Sugar-making cells are found in large numbers in the shoot, and only rarely in roots. Many of the other kinds of cells are found in both.

What differences in organisation are there?

The chief points are that:

(A) shoots consist of **stems** and **leaves,** with buds and branches formed at the points where the leaf is attached to the stem (the *node*); ***whereas***

(B) roots are variations on one kind of structure, and can branch more or less anywhere.

In addition, shoots have a much thicker waxy covering on surface cells (C 10), which reduces water loss.

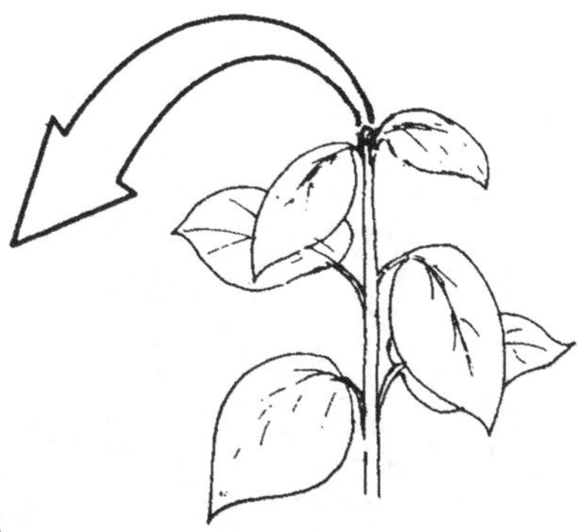

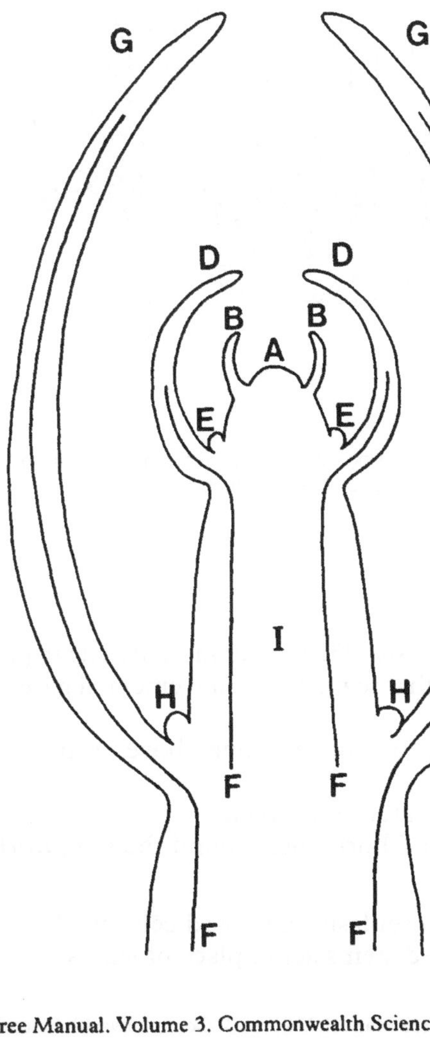

Section through a young shoot.

A — shoot tip. B — first pair of leaf projections. D — third pair of young leaves *(the second [C] cannot be seen)*. E — third pair of young buds. F — conducting cells. G — fifth pair of leaves expanding. H — fifth pair of buds. I — third and fourth internodes extending.

How do stems and leaves grow?

They are both formed at shoot tips, which have zones of:

(1) **actively dividing cells** at the tip, which from time to time form small projections that become leaves;

(2) **dividing and elongating cells** behind the tip, which form the new *internodes* of the stem, extending and pushing the shoot tip farther away from the roots, and also expanding the leaf; ***and***

(3) **specialising cells** (including sugar- and water-conducting cells) in the leaves and the parts of the young stems that lie behind zone (2) and have ceased to extend.

What makes stems and leaves different?

Although both of them arise from the same growing point, they differ because:

(A) **leaves** are temporary organs that are typically predetermined as to their maximum size, and can seldom form a new stem; ***while***

(B) **stems** can usually continue to grow indefinitely, producing more stems and leaves. Some become progressively thicker, forming the trunk and major structural branches.

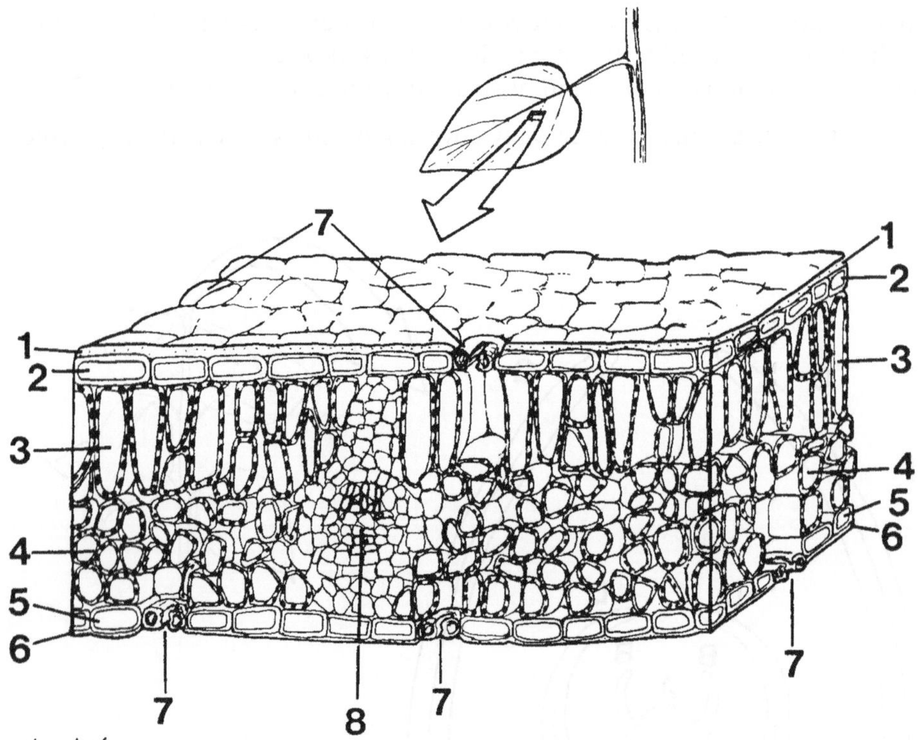

Section through a leaf.

1 – thick waxy layer. 2 – upper surface cell. 3 – upper sugar-making cell. 4 – lower sugar-making cell. 5 – lower surface cell. 6 – thin waxy layer. 7 – guard cells and holes. 8 – conducting cells.

Don't they also differ in what they do?

Yes, that is so. Generally:

(1) leaves are primarily organs of photosynthesis. They contain many sugar-producing cells and also special surface cells *(guard cells)* that regulate the movement of gases through the leaf surface (C 10, C 13); ***whereas***

(2) stems have several functions, including providing a strong above-ground structure that:

(a) displays a succession of leaves, and later on flowers;

(b) conducts water, nutrients, sugars and hormones around the tree; ***and***

(c) stores starch and other substances.

However, most young stems also contain some chloroplasts and guard cells, while in a few tree species, such as *Acacia mangium,* short flattened stems act in place of leaves.

What happens to the leaf projections at the shoot tips of trees?

They may:

(1) develop into full-sized leaves at once, by continued cell division, expansion and specialisation;
(2) grow a little and then remain within a bud for a time, and then expand rapidly when the bud *flushes* later on;
(3) become a small bud scale around an inactive bud; ***or***
(4) never get much bigger, and soon fall off.

What about the shape of leaves?

This varies greatly between different species, to some extent from one tree to another of the same species, and even within a single tree. For example, leaves may have:

(1) **stalks** that are long, short or absent altogether;
(2) **blades** which are divided up into separate leaflets, or not;
(3) **margins** that are deeply lobed, finely toothed, wavy or uncut;
(4) **surfaces** which are flat or undulating, hairy or smooth; ***and***
(5) **texture** that is tough and hard, or thin and delicate.

Many features of leaf shape are useful in identifying one tree species from another.

How big can leaves get?

Their final length can vary between a few millimetres (for instance in *Casuarina* and *Cupressus*) and several metres (for example in palms). Widths range from about 1 mm to 1 m.
In mountains, tree leaves are usually smaller than at lower elevations, and the same is often true in drier compared with wetter environments.

Which features of leaves are the most important?

The **number of leaves** and the **leaf area** are usually key points, whether for a single leaf, a tree crown or an entire stand.

What determines leaf area?

Besides the genetic potential of the individual tree (C 5), many other factors influence the number of leaves formed and how big they grow. These often include:

(A) ***in the aerial environment*** (D 11 in Manual 4):
  (1) the air temperature, during the night as well as the day;
  (2) how much light falls on the leaves during an average day;
  (3) the *quality* of this light, and the day-length;
  (4) the drying power of the air (determined by temperature, humidity and wind speed);
  (5) competition with the shoots of other plants;

(B) ***in the soil:***
  (1) the physical and chemical nature of soil components (D 12-13 in Manual 4);
  (2) soil temperature;
  (3) the supply of water to the tree, compared with the amount the shoots are losing (C 13, C 43);
  (4) the availability and balance of nutrients (C 14, C 33);
  (5) the presence of close associations between micro-organisms and the tree's root system (C 30-32);
  (6) competition with the roots of other trees and weeds (C 44); ***and***

(C) ***within itself:***
  (1) how big the crown is, whether the leaf is on the main stem, a large branch or a small twig, and where it is positioned on the particular shoot;
  (2) the interplay (past and present) between root and shoot, and between vegetative and reproductive activity; ***and***
  (3) the presence of stored reserves of sugars and nutrients.

Do I want the leaves to be as big as possible?

Yes, in a well-established tree, growing in the ground; ***but***
No, in a young tree in the nursery (C 34).

How long do leaves live?

Usually between 3 and 15 months, although in a few tree species they may last for several years. In a rapidly growing nursery tree, some of the first-formed leaves may live for only a few weeks.

What happens to their contents when they die?

(a) Normally, when old leaves start to change colour, some of the sugars and nutrients in them are transported out into the stem. Then the leaf is actively cut off by the tree and falls to the ground, where it adds to the litter (D 13 in Manual 4); ***but***
(b) Under stress conditions, leaves may just wilt and shrivel up, often staying hanging on the tree.

In a few species, such as some palms, it is normal for the bottoms of dead fronds to remain attached to the trunk.

Why do some trees naturally lose all their leaves at once?

This is a common feature of:

(1) many trees and shrubs in savanna and dry forest areas;
(2) some of the taller trees in humid forests.

Deciduous trees shed all their leaves well **before** they produce any new ones, so that the tree stands leafless for weeks or months;
Leaf-exchanging trees shed their leaves at about **the same time** that they produce new ones, so the tree may be leafless for a short time only;
Evergreen trees never stand leafless, because they either:

(a) shed their old leaves well **after** new ones are formed; ***or***
(b) grow **continuously,** producing new leaves and shedding old ones all the time.

Is it drought that makes leaves fall off?

Although many leaves may fall during the early part of the dry season, there are probably several other factors at work, such as:

(1) other small changes in the environment;
(2) the leaves may be near the end of their natural life-span, with other more recent leaves 'taking over'; ***or perhaps***
(3) monkeys or a swarm of insects might have eaten most of the leaves.

Poor watering, rough handling or sudden changes in the amount of light reaching nursery trees can sometimes be followed by a lot of leaf shedding (C 40-43, C 47).

So shoots don't necessarily grow all the time?

Shoot growth is **continuous** in young trees of a considerable number of species; ***but***
Some seedlings and most saplings and older trees show **periodic flushing,** followed by one to eleven months in which no new leaves are expanding or stems extending.

Is it rainfall that starts shoots growing again?

Perhaps yes in a few species, especially in dry areas; ***but***
Probably no in many others, because flushing commonly occurs before the first rains.

Do all the buds grow into branches?

No. Some buds may:

(1) never grow any bigger;
(2) remain small and inactive unless the shoot is damaged;
(3) turn into flower buds (Manual 2);
(4) become *dwarf shoots* of limited growth potential. For example, the needle-like leaves of *Pinus* are produced on a tiny shoot, while the leafy, budless branchlets of *Taxodium* are shed like leaves.

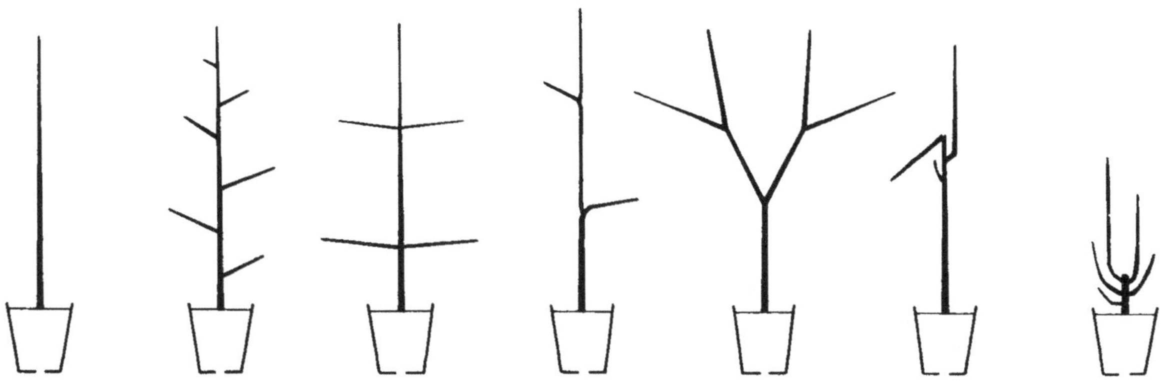

Common types of branching.

Which buds become branches?

Trees generally have regular patterns of branching that produce a growth habit and crown shape that are characteristic of the species, and sometimes of the individual clone (A 11-13 in Manual 1).

Does this matter much?

Yes it does, because different branching habits and crown shapes are best for trees needed for shade, shelter, wood production or fruits.

For instance, straight stems are valued for timber, pulpwood and poles (C 36-38 in Manual 4), but undomesticated tree species (C 5) often contain stems that are crooked, forked or even grow as multiple stems. Some may have large branches that leave big *knots* in the wood, reducing its strength and value.

How do stems get thicker?

By division of a cylindrical layer of cells lying nearer the outer surface than in roots (C 11). The cells it produces on the inside become specialised as wood cells (water-conducting, strengthening or storage), while those formed to the outside are sugar-conducting or strengthening cells (C 10).

Will all these cells stay alive?

Dividing layers and storage cells may live for many years, and sugar-conducting cells for a shorter time, but many of the other kinds die quickly. The outer part of the wood *(sapwood)* contains some living cells, but in the inner *heartwood* they are all dead, though they generally remain strong.

What happens to the surface of the tree when it grows?

When a **leaf** falls off, its *scar* has usually been already protected.
As a **stem** grows in thickness, a second layer of dividing cells is formed nearer its surface. This produces the bark, and the original surface cells die. The bark may become smooth, ridged, scaly or peeling, but looser areas remain that allow some continued gas exchange.

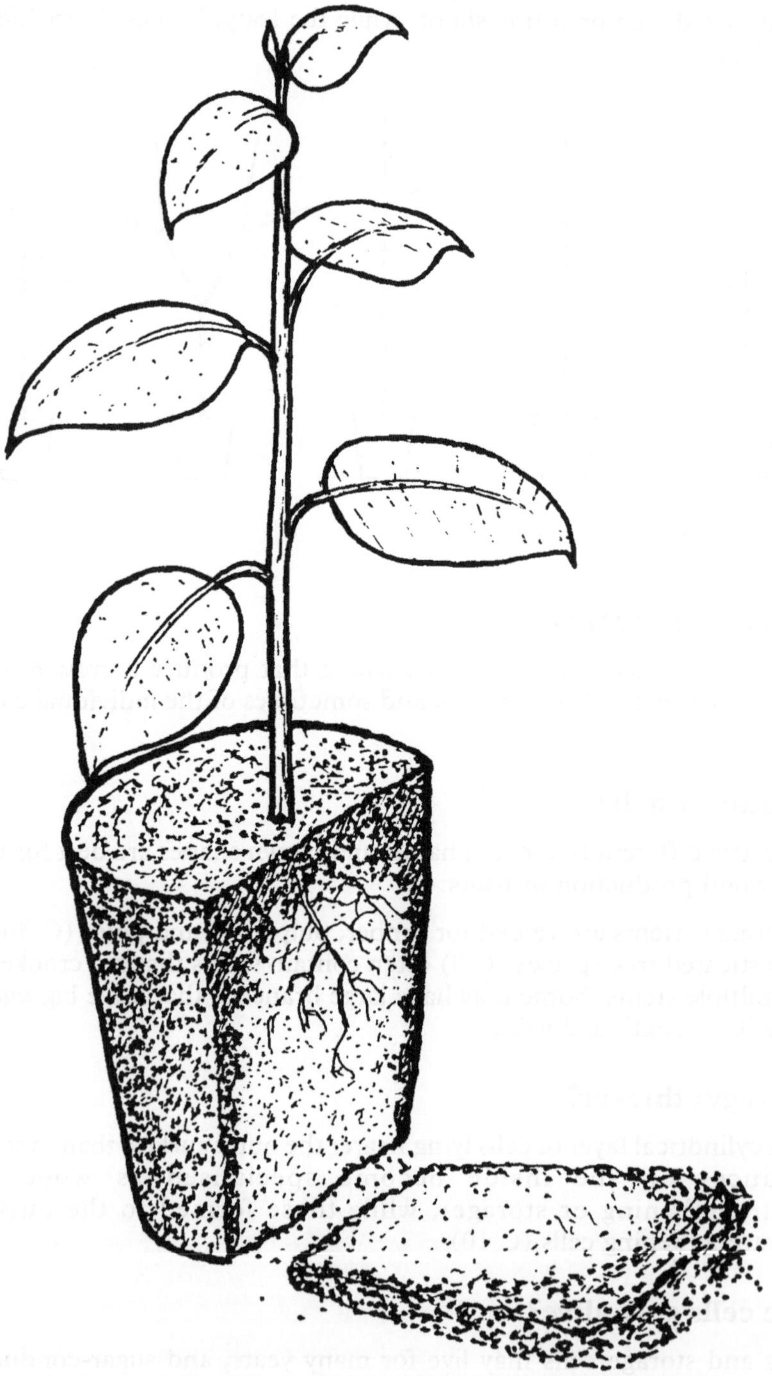

Cutting a test plant shows that water did not reach all the soil.

General principles of tree growth C 13
- *maintaining the water balance*

What is meant by the water balance?

Trees contain large quantities of water, but if they **lose** more than they **gain,** there may soon be problems.

How much water is there in a tree?

A considerable proportion of it consists of water. For example:

(a) a thin leaf might be about 90% water by weight; ***and***
(b) an expanding root cell could contain as much as 95% of water.

Is it important that there should be so much?

Yes, because:

(1) many of the activities of living cells (C 10, C 14) are carried out in very dilute watery solutions;
(2) water is needed for the new cells in young stems, leaves and roots to elongate;
(3) plenty of water is required to maintain considerable pressure inside living cells (C 10), giving support to expanded leaves, young stems and flowers;
(4) trees that are in leaf generally lose large amounts in a single day, and need some water in reserve.

What happens if a tree runs short of water?

It is under *water stress* (C 41). Moderate, temporary water stress is normal, but a lack of substantial amounts of water for longer periods can lead to premature leaf-shedding, shoot die-back or death of the whole tree, unless it:

(a) is of a species and genetic origin that comes from a drought-prone area;
(b) has had time to form a well-established root system (C 11, C 42); ***and***
(c) has been *hardened* before being exposed to full light and drying winds (C 47).

And if not, it wilts?

If nursery trees wilt, it means that the water shortage is decreasing the pressure in the cells. This should be avoided, as they are already under severe stress, leading to a check in growth, damage or even death.
Note: young trees with thick, leathery leaves and tough stems may not wilt although suffering from pronounced water stress.

Why do trees in leaf lose such a lot of water?

Because it *evaporates* so easily from leaves, especially from their moist **internal surfaces.** Perhaps only around 10% is lost from the outer surfaces of leaves, stems, aerial roots, flowers and fruits, because:

(a) the surface cells are covered by a waxy substance that restricts evaporation;
(b) thicker stems generally develop bark (C 12), which has a similar effect.

Less than 1% is actually used up in photosynthesis and other chemical reactions.

How does such a lot come from inside the leaves?

Because there are:

(1) a series of interconnected air-spaces inside the leaves, filled with moist air;
(2) a lot of sugar-producing cells exposed, which lose water easily; ***and***
(3) many small holes between the surface cells (C 10), which let carbon dioxide enter freely, but also let water vapour escape rapidly.

The bigger the arrow, the more water is evaporating.

Can they be closed?

Yes, each small hole can be opened or closed by two special *guard cells.* Generally the holes are ***open*** during daylight hours, ***but closed:***

(a) during the night, when photosynthesis cannot occur; ***or***

(b) if the tree is suffering from pronounced water stress.

It is also common for many of the holes to close around midday during sunny weather.

But isn't that when sugar production could be fastest?

Yes it is. Light levels are often greatest then, but with higher air temperatures evaporation will also be very rapid. A balance is struck by the plant, capturing carbon dioxide but stopping excessive water loss.

In practice, unshaded or lightly shaded leaves in the nursery can receive plenty of light for photosynthesis from about 0800-1200 and 1500-1700.

What influences how much water is lost by a tree?

(1) The amount of sunshine reaching it;
(2) The temperature and humidity of the air;
(3) How strong any wind or air currents are;
(4) How many leaves there are on the tree, whether they are hairy or not, the angle they are held and whether they are folded or rolled;
(5) Whether most of the small holes in them are open or closed.

How quickly can water be lost from leaves?

If you break off a leaf and keep it in the sun, it may be only a minute or two before it wilts. Water has been lost so rapidly that most cell activities will already have been affected. Unless the water balance is quickly restored, the leaf will soon die.

How much water does that mean in a day?

Each attached leaf could lose many times its own weight in water in a single hour. So a small tree may lose litres a day, and a big emergent tree in the canopy of a tropical forest might be losing hundreds of litres an hour.

Where does all this water come from?

It has to come from the soil, by way of the root system, trunk, branches and twigs.

However does it get up a tall tree?

Most of the water travels in the dead water-conducting cells (C 10), which are like long, miniature pipes.

Aren't they too small to carry all that water?

No, because there are a lot of them, all the way from the absorbing zones in young roots (C 11) to the small veins in the leaves.
Their diameter is usually much less than 0.5 mm, but this and their wall structure help these tiny water columns to travel up and not break.

How do the roots pick up so much water from the soil?

A well-developed root system has a very large *surface area* of fine, growing roots in contact with moist soil particles. Water enters the root easily because:

(a) the **cell walls** of the surface cells of the root offer little or no barrier to its movement; ***and***

(b) their **cell membranes** tend to pass more water in than out, because of the dilute solution inside them.

The surface area of the root system may be enlarged still further by:

(c) thousands of fungal threads from mycorrhizal associations (C 30-31); ***and sometimes***

(d) root hairs, which are temporary extensions of the root surface cells.

What affects how fast it comes in?

The rate at which water is taken up by the root system depends primarily on how much water is being lost from the shoots of the tree, which draws it up from the roots. However:

(1) water loss may exceed uptake during the day-time, in which case the trunk of a large tree can shrink measurably, as water reserves inside it are depleted; ***but***

(2) water uptake continues during the night, often restoring these lost water reserves by 0600 hrs, allowing the trunk to return to full size.

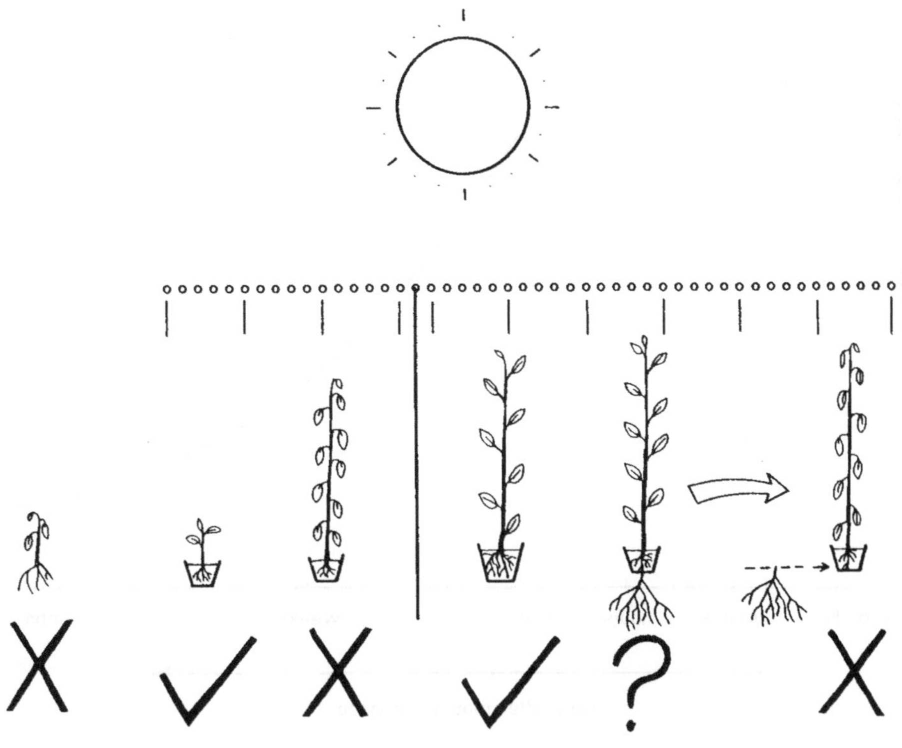

But is there enough water in the soil?

For an established tree there may be, but its root system must be able to:

(a) keep growing and branching into new soil, maintaining a large surface area of absorbing roots; ***and***

(b) take up water from the soil even when it is getting dry.

For a young tree in a container, a good root system (C 4) and adequate watering (C 43) are clearly vital. If the potting soil becomes dry, water may still be lost rapidly unless most of the small holes in the leaves have closed.

Is there any way of saving young trees that are wilting severely?

(A) Reduce water loss immediately by shading plants growing in beds, and moving containers to a humid, shady and protected place (ideally a poly-propagator, see A 31 in Manual 1); ***and***
(B) Water the plants if the soil is dry.

Note: *do* ***not*** *water if the soil is moist, because it will not help, and it could cause harm by waterlogging the soil (C 11), and encouraging root diseases (C 45).*

Can shoots take up any water directly?

Generally no, because rainfall does not normally enter the small holes in the leaves. Occasionally they might do so, for example when the soil is very dry, and:

(a) mist or dew moisten the shoots;

(b) the leaf surface is specially adapted, or aerial roots are formed, making above-ground water absorption possible.

Although leaves and young stems depend on rain taken up by the roots, if water fills their internal air spaces they cannot function properly.

What about trees which have shed their leaves naturally?

These lose much less water than leafy plants, and may even contain sap under pressure. They should be watered sparingly (C 43), but not allowed to dry up.

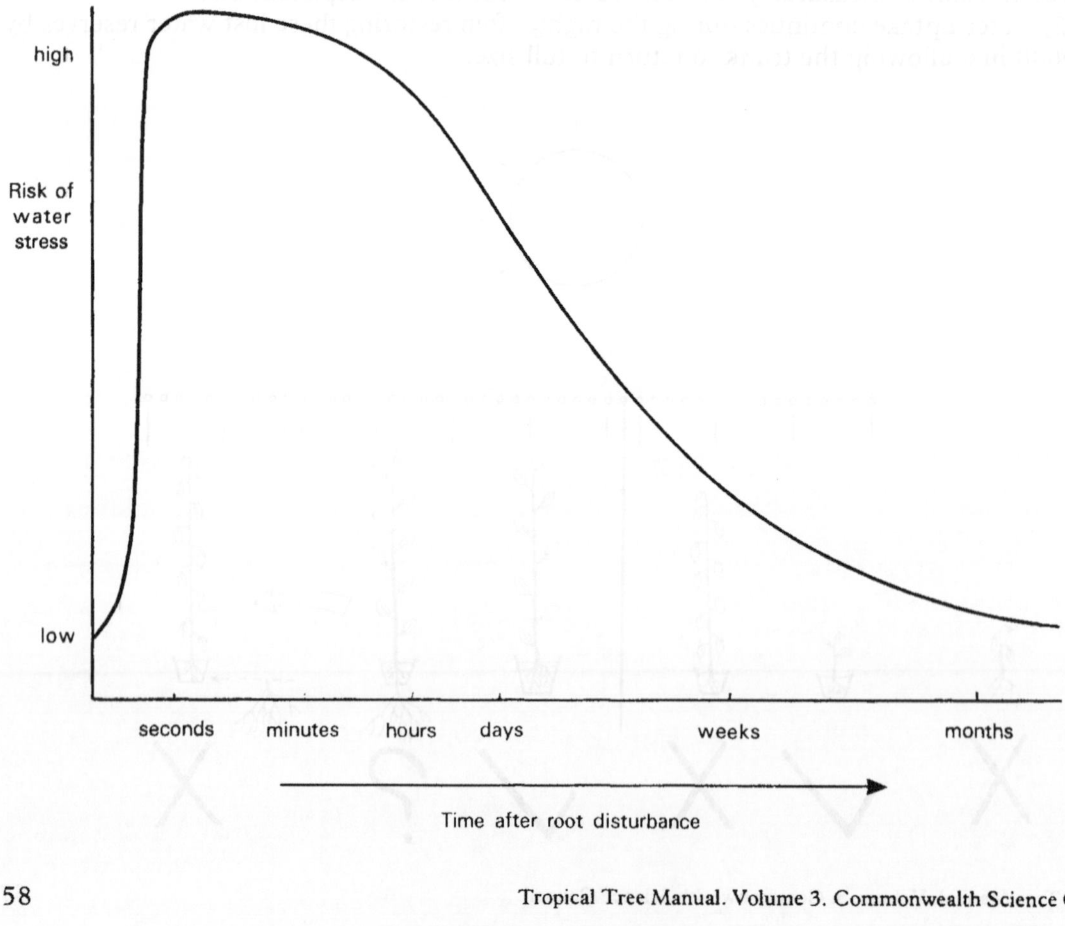

Are there some other practical guide-lines about water balance?

(1) Young trees, even of hardy species and local origin, are very liable to run short of water when their root systems are disturbed, so damage needs to be kept to a minimum when they are potted up (C 42) or moved to different conditions (C 41).
(2) Unrooted and newly rooted cuttings (Manual 1) and germinating seedlings (Manual 2) require special care because they are particularly liable to water stress.
(3) Most tree species need shading (C 41) at some stages while in the nursery, and careful watering (C 43) is needed by all young trees, especially in containers (C 6).
(4) Growing a good root system (C 4, C 11) is more important than having a big shoot (C 12) on a nursery tree. Similarly, large leaves are undesirable on planting stock (C 34).
(5) Young trees need 'hardening' before going to the planting site, so that the newly planted tree can grow "from a position of strength" (C 47; and Manual 5).
(6) The chief function of a tree nursery is to produce planting stock that can maintain its water balance when planted out.

Other suggestions for maintaining the water balance are found elsewhere in the Manuals.

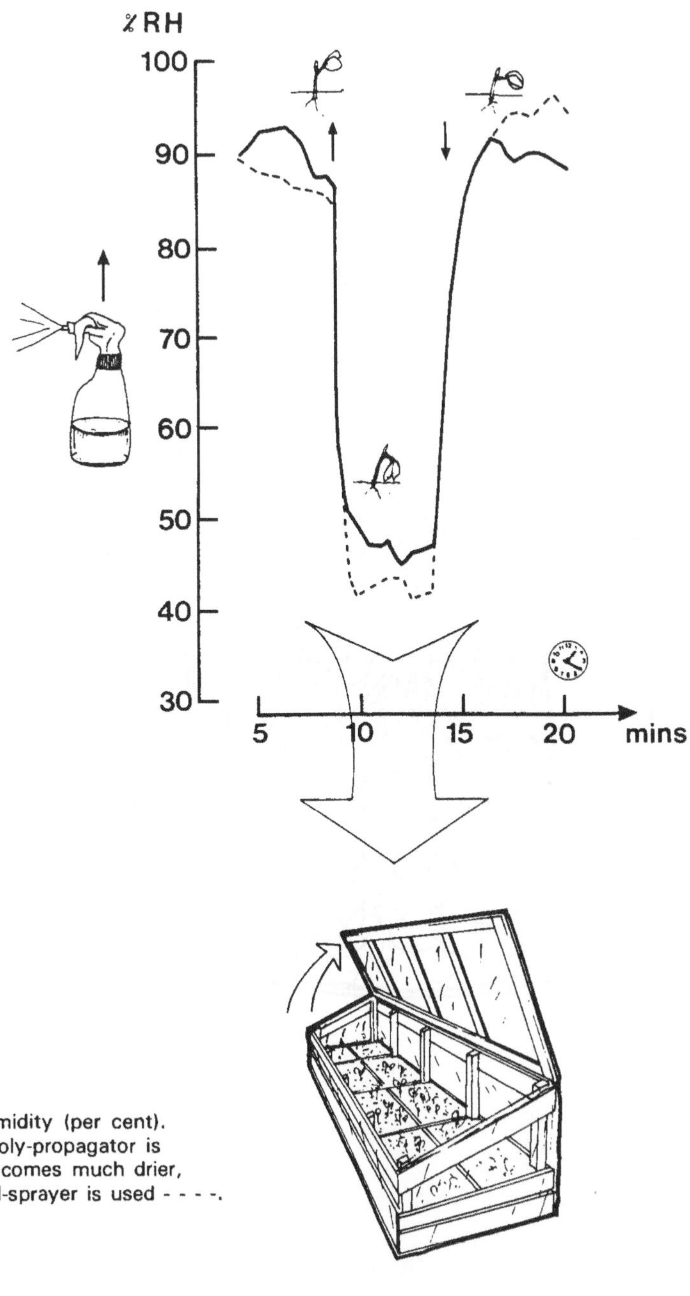

%RH = relative humidity (per cent). Immediately the poly-propagator is opened, the air becomes much drier, even when a hand-sprayer is used - - - -.

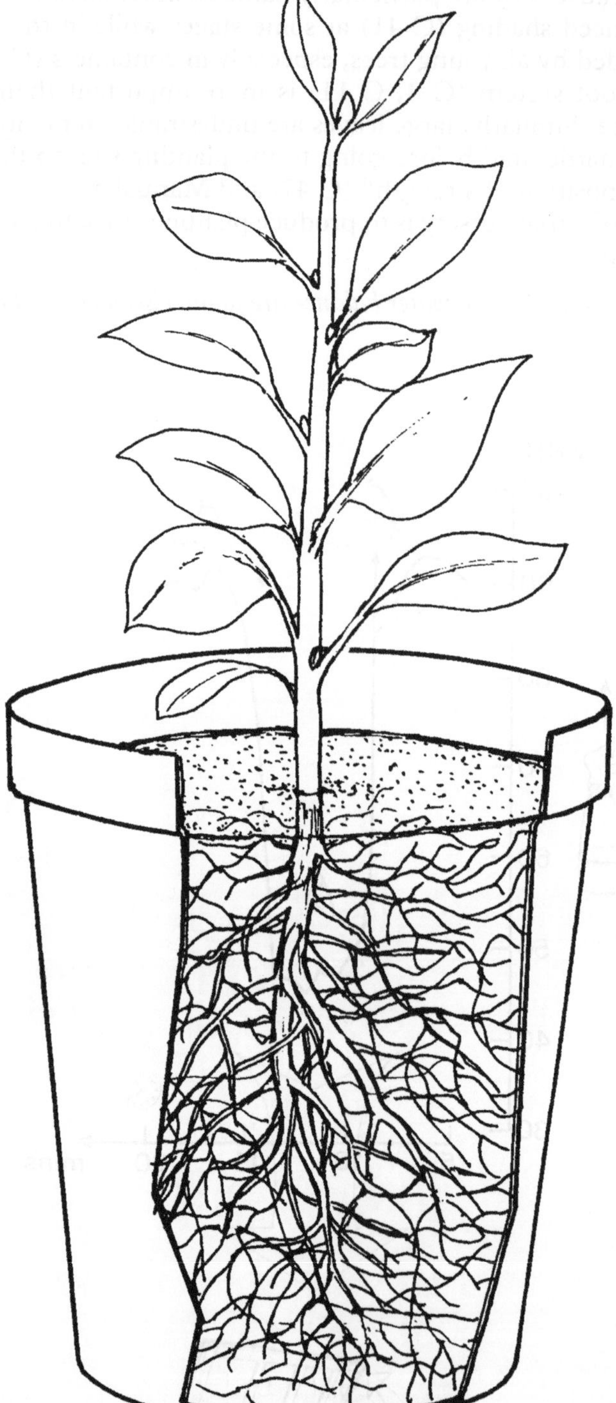

General principles of tree growth C 14
- *absorbing, making and moving substances*

What kind of substances does the tree absorb?

(A) **Water,** mainly taken up by the roots from the soil (C 11, C 13);
(B) **Nutrients,** also chiefly from the soil; ***and***
(C) **Carbon dioxide,** as a gas which passes in the air into the leaves (C 10).

Some **pollutants** can also be absorbed by the shoots of trees (D 16, D 26 in Manual 4).

What are nutrients?

Simple chemical substances that are important to or essential for the growth of all living organisms (D 13 in Manual 4).

Which are the main nutrients?

(1) **Nitrogen (N),** occurring in soluble compounds such as urea, nitrates and ammonium salts, and in a less easily dissolved form within organic matter;
(2) **Phosphorus (P),** usually in an insoluble form, but more soluble when it occurs as various kinds of phosphates; ***and***
(3) **Potassium (K),** usually as fairly soluble salts.

Relatively large amounts of these three nutrients are required for good growth of both young nursery plants (C 33-34) and older trees (D 13 in Manual 4).

Are other nutrients needed as well?

(a) Moderate quantities of substances containing **calcium (Ca), magnesium (Mg)** and **sulphur (S);** ***and***
(b) Very small amounts of *micronutrients,* from chemical compounds containing **boron (B), chlorine (Cl), cobalt (Co), copper (Cu), iron (Fe), manganese (Mn), molybdenum (Mo)** and **zinc (Zn).**

Where do all these chemicals come from?

(1) **Litter:** nutrients are released by the decay of organic matter:
- (a) on the surface of and within the topsoil (D 13); ***and***
- (b) when used as an important component of potting mixtures (C 6);

(2) **Soil particles:** some nutrients become available as any stones and gravel (D 12) are very gradually broken down into smaller sizes by *weathering* ;
(3) **Air:** a few nutrients come from gases in the air, and from dust and other tiny particles that can settle or be brought down in rainfall.

Which is the most important source?

Under trees, the **litter** is usually by far the most important. Most nutrients are very efficiently recycled in natural woodland ecosystems, so that they are used over and over again, with only very small losses in *run-off* and *leaching* (D 13).

What happens when the trees are cut down?

Nutrient supply generally becomes a serious problem, because:
- (a) most nutrients are very easily **lost** from land that is bare, farmed without shade trees or planted with a single species of tree (Manual 4);
- (b) many tropical soils contain **limited supplies** of nutrients; ***while***
- (c) **growth** of plants can be **greatly retarded** if even one kind of nutrient is in short supply.

So recycling of nutrients is one of several important reasons for having plenty of different kinds of trees (C 1; and D 30 and D 53 in Manual 4) in managed ecosystems.

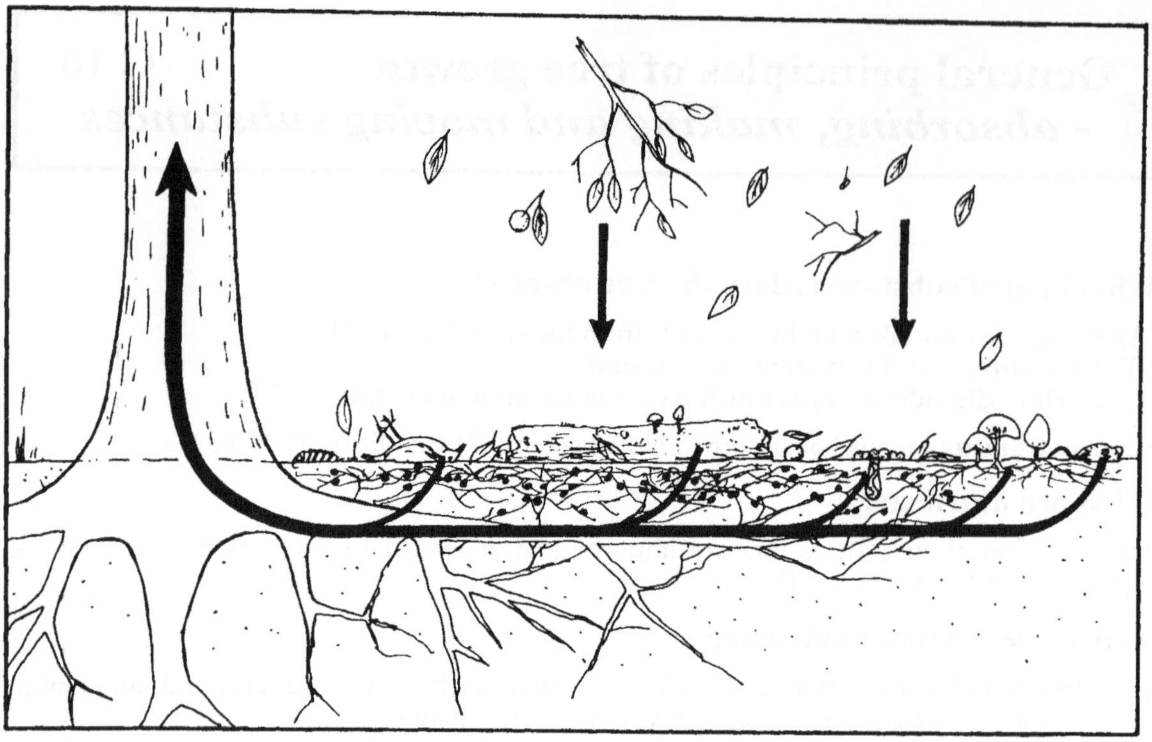

Recycling of nutrients.

But does this apply to young trees in the nursery?

Young nursery trees start off with a supply of nutrients stored in the seed, or present in the cutting when it is taken. But sooner or later they depend for nutrients on what is available in the potting mixture or nursery bed.

What about adding fertilisers?

If a *balanced* fertiliser (C 33-34) is added to a poor soil, the young trees may be able to take up more nutrients and grow better, unless:

(a) these are rapidly washed out of the pots, or *leached* too deep for the roots;
(b) the soil is too acid or too alkaline (C 6);
(c) there is still a lack of one or more individual nutrients; ***or***
(d) there was plenty of a particular nutrient already in the soil, and there is now too much, damaging the roots or hampering the uptake of another nutrient.

It is generally better to concentrate on choosing good nursery soils (C 23) and preparing favourable potting mixes (C 6).
Use cheap and readily available materials like composts and wastes (C 33), and avoid expensive fertilisers except when they are specially needed.

How do nutrients actually get into the tree?

(1) They must dissolve in water, even if sparingly;
(2) The solution enters the network of cell walls in the outer parts of the absorbing regions of fine roots (C 11);
(3) Specific nutrients are actively transferred across the cell membrane (C 10) into the living cell, while others are excluded or absorbed in smaller quantities; ***and***
(4) They pass an inner boundary layer and can move to the rest of the tree.

Small amounts of nutrients can also be absorbed by leaves, if for example a foliar feed is sprayed on to them.

What about mycorrhizas?

Yes, these are very important in the uptake of nutrients (C 30-31; and D 32 in Manual 4), because the fungal threads can:

(a) form a very extensive network that captures and recycles a lot of nutrients;
(b) often release P and other nutrients that are in an insoluble form in the soil; ***and***
(c) pass nutrients into the tree's root system.

How about nitrogen-fixing trees?

These form nodules with closely associated micro-organisms that have the unique capacity to fix N from the air, where it is very abundant, into nutrient form (C 32 and D 32).

Are sugars also nutrients?

No, they are more complex **organic** substances that green plants ***manufacture*** from simple chemicals containing **carbon (C),** using the energy of sunlight (C 10).

Does that include carbon dioxide?

Yes, the energy that is captured during photosynthesis is used to make sugars by adding more carbon to organic acids, which also contain **hydrogen (H)** and **oxygen (O).** Some of the latter is released into the atmosphere.

What other kinds of substances do trees manufacture?

Amongst the most important are those used in:

(1) making **cell membranes** (C 10);
(2) building **cell walls,** including *cellulose* and various substances that strengthen and join cells together;
(3) copying the **genetic instructions** during cell division;
(4) providing nitrogen-containing building blocks for making the ***enzymes*** **(proteins)** that allow specific chemical reactions to take place rapidly;
(5) creating the phosphate-containing substances that allow **easy transfer of chemical energy** to drive chemical reactions, including photosynthesis;
(6) storing **starch, fats** and other substances for later use.

Do trees make anything else?

Yes, a very large number of different chemicals. Some have been widely studied, while many others are yet to be identified. For example, some may:

(a) discourage most kinds of animals from eating the leaves or bark (D 10 in Manual 4);
(b) give an attractive odour to flowers or flavour to edible fruits (D 33);
(c) contribute to the strength or durability of timber (D 36);
(d) provide useful substances that can be used for raw materials (D 37), medicines (D 33) or veterinary purposes (D 39).

How do these substances move around in the tree?

Those that do so will move:

(A) from one living cell to the next through tiny connections;
(B) along the fine tubes formed by the elongated water- and sugar-conducting cells.

Some substances are formed, broken down and reformed many times over.
However, others are made within a cell and never move, becoming part of the permanent tissues of the tree.

Where do sugars move?

From the sugar-making cells in the leaf and stem to neighbouring cells, and then through the sugar-conducting cells to all other cells in the shoot and the root system.

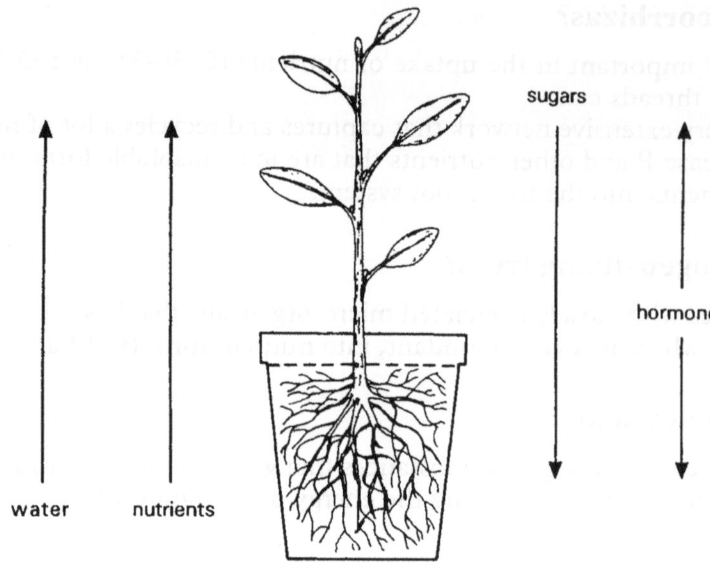

What are they used for by the tree?

Besides storage, sugars are broken down *(respired)* (C 10) in each living cell, releasing energy for:

(a) its own maintenance and manufacture of chemicals;
(b) cell division and expansion in the growing zones of the tree, and repair of damaged parts; ***and***
(c) driving the active uptake of nutrients and the transport of organic substances.

How do nutrients move?

Simple nutrients, once through the inner boundary layer of an absorbing root, travel mainly through the water-conducting cells to other parts of the tree.
Organic nutrients, already partly built up, may pass either by the same route or through the sugar-conducting cells.

How is all this controlled?

In at least 3 ways:

(A) through the **genetic instructions** contained in the nucleus of each cell (C 10), which accurately control the production of enzymes and thus of other substances;
(B) by the **position within the plant** that a cell occupies, which narrows the wide potential for activities it possesses; ***and***
(C) by **plant growth hormones** which are manufactured in one part of the tree and often move to another.

For example, when a leafy stem cutting is detached from a stockplant (Manual 1), the natural auxins (produced by the shoot tip and moving towards the base) can help to regenerate a complete plant by stimulating cells there to form new root tips. Synthetic auxins applied to the base can often increase this process (A 40).

What else do plant hormones do?

Various hormones appear to help keep a balance between the two parts. For example:

(1) poor root systems generally lead to poor shoot growth;
(2) very rapid shoot growth may coincide with reduced root activity, followed by the reverse;
(3) poor shoot and root growth may follow heavy fruiting.

Are hormones well understood?

Although control over the growth and development of large trees is a central part of biology, there is still much to be learnt. Much more research is needed (C 7, C 15).

General principles of tree growth C 15
- *experiments with young trees in pots*

What experiments can be done in pots?

Many different kinds. It is often best to concentrate on those which are:
(1) fairly straightforward to do; ***and***
(2) likely to help increase understanding of tree growth and to improve nursery practice.

Are there some examples?

(A) Discovering more about the various **factors that affect tree growth** (C 10-14, C 62);
(B) Comparing different **growing conditions** for a batch of trees;
(C) Seeking to resolve **nursery problems** that have been experienced (C 3, C 60-61);
(D) Testing **tree species** and **genetic origins** (C 5) that are unfamiliar;
(E) Trying out a new **technique** or **timing.**

What is the main aim of basic studies of tree growth?

To add a bit to our understanding of some of the complex **interactions** (C 69-K) between:
(a) the genetic potential of the individual tree (C 5);
(b) internal factors dependent on the size and age of the tree (Manual 2);
(c) the non-living part of the local environment;
(d) competition and collaboration with other plants and micro-organisms;
(e) effects of animals, especially humans.

How would I discover better growing conditions?

By using the most successful local techniques, and then:
(A) Setting up an experiment, in which you could for instance compare:
(1) different types or sizes of containers (C 6);
(2) contrasting potting mixes;
(3) soils inoculated for mycorrhizas (C 30-31) or nitrogen-fixing nodules (C 32), compared with uninoculated control trees;
(4) differing watering regimes (C 43);
(5) degrees of shading, or rates of reducing it (C 41);
(B) Trying out the most successful treatments on some spare batches of trees; ***and then***
(C) If appropriate, modifying current nursery methods accordingly.

What about problem-solving?

Some nursery problems can be tackled by:
(a) thinking about the general principles that control the growth of trees (C 10-14), and spotting which conditions might be unfavourable;
(b) using a check-list (D 60); ***or***
(c) checking with a more experienced grower or with research workers (C 53; and D 5 in Manual 4).

Other problems may need formal or informal experiments to find out what might be wrong.

Is it possible to test out a new tree species in pots?

Yes, this is often a useful first step, ***before*** trying them out in field trials (D 29 in Manual 4). One might discover, for instance:
(1) if the new trees can thrive under nursery conditions that suit other species, or whether they may need special environments (C 48);
(2) what their growth rates are like; ***and***
(3) if they are particularly liable to damage by stress (C 41) or from insect pests or diseases (C 45).

Before writing off the species as a failure, remember that it could be the provenance, seed-lot or clones used (C 5) that were unsatisfactory.

How would I set about doing an experiment?

(a) Grow enough young trees to a suitable size under uniform conditions in containers (C 6-7), allowing for some spares;
(b) Decide on the particular topic for the experiment, and what treatments to apply (D 6 in Manual 4);
(c) Get together all the materials and tools you are going to need (C 24, C 51, C 61-B), including labels, pencil and paper;
(d) Choose enough uniform, undamaged plants out of the batch to use, or *grade* them if numbers are short (A 45 in Manual 1), and allot them to treatments without bias;
(e) Apply the treatments, if possible all on the same day, label the young trees clearly, and make a simple record of the date, treatments, number of plants and their origin (C 54);
(f) Arrange the containers in a suitable layout (C 62-F; and D 55 in Manual 4);
(g) Look after the trees well (C 48);
(h) Note down (with the date) any differences you observe, and assess the experiment at appropriate intervals (C 55, C 67-68); ***and***
(i) Draw valid conclusions from the results, analysing them where appropriate (C 69).

Supposing I want to compare pots of different sizes?

They should preferably be:

(1) all made of the same kind of material, with a similar shape; ***and***
(2) filled with the same batch of well-mixed potting soil; ***but***
(3) sufficiently different in size for any effects to have a chance of showing up.

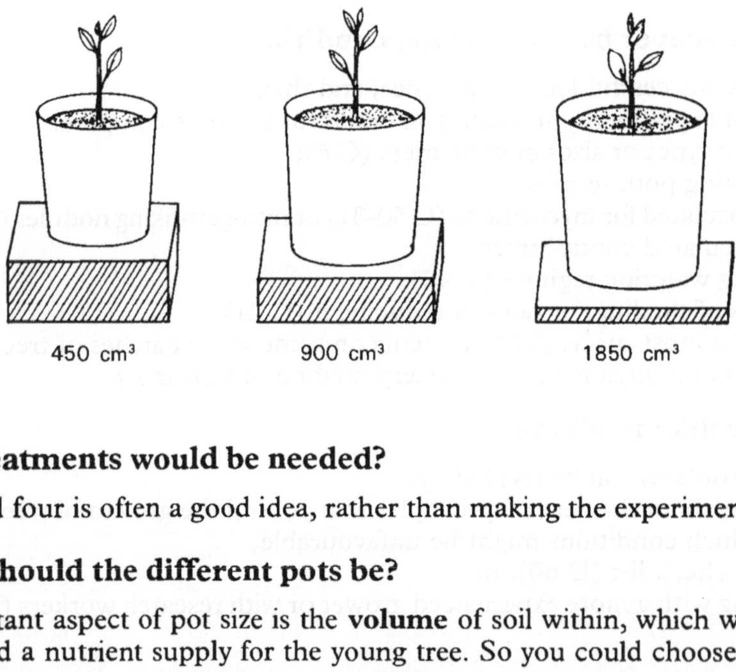

How many treatments would be needed?

Between two and four is often a good idea, rather than making the experiment too large.

And how big should the different pots be?

The most important aspect of pot size is the **volume** of soil within, which will provide both rooting space and a nutrient supply for the young tree. So you could choose convenient pot sizes that:

(a) range either side of a container of medium size;
(b) involve two- to three-fold differences between treatments.

What would that mean in practice?

If, for instance, you chose tapered pots where the diameters at the level of the soil at the top were 9.5, 12.25 and 15.5 cm, this would give soil volumes for the root systems of about 450, 900 and 1850 cubic centimetres. The medium-sized pots would hold about twice the soil of the small ones; and similarly the large containers would have about double the contents of the medium-sized pots.

How can I check the volume of my containers?

See sheet C 63-D for this.

Should all the containers have the same number of holes?

No, the medium-sized pot would need to have more holes than the small one, so that excess water drained off in a similar way. The large containers would need even more holes to maintain comparable conditions of soil moisture.

How would I set about grading the young trees?

Supposing you have decided to have 25 *replicates* for each of the 3 treatments:

(a) first arrange the 75 plants in *triplets* of similar height, branching habit, vigour or other features;

(b) next *randomly* assign the 3 plants of the first triplet to each of the 3 treatments. (You could do this by withdrawing small, similar-sized, numbered pebbles in turn from a bag, after mixing them together); ***and***

(c) repeat for the other 24 triplets.

What about applying the treatments?

Some important points to plan for are:

(1) Reduce to a minimum the time that root systems are exposed to the air;

(2) Pot up the whole experiment on the same day, or if this is not possible, complete a certain number of triplets, not treatments; ***and***

(3) Try as far as possible to keep conditions the same for all the young trees, except for the pot size. Avoid favouring any particular treatment, as this would *bias* the results of the experiment, perhaps giving you false ideas about the most suitable pot size. For example, try not to damage roots more when putting them into small pots.

How about labelling and records?

The best way is to:

(A) label each plant with numbers and/or letters that identify it uniquely; ***and***

(B) note down the details of the experiment straightaway, including the date and the origin of the plants which were used (C 54).

Are individual labels really necessary?

Although in this experiment you can easily see which treatment is which, remember that:

(1) knowing which trees were in which triplets can improve the analysis of the results (C 69-F, I);

(2) labelling trees individually will mean that their performance can continue to be followed after they have been planted out.

Note: experiments using clonal plants (Manual 1) can give greater precision than those with seedlings, provided that the trees of each clone are allocated equally to the 3 treatments. In such cases you might perhaps write the treatment details on the label bearing the clone number.

Where should the pots be kept?

(a) In as uniform conditions of light as possible; ***and***

(b) Where you can water (C 43) and check the experiment regularly (C 40, C 48).

It may be useful to put an extra line of similar trees in pots around the experiment, so that none of the treated trees are 'edge plants' subject to different conditions.

How should the experiment be laid out?

In such a way that no treatment (or genetic origin) is specially favoured. To avoid bias it is often recommended that a completely random arrangement of the pots should be used. However, you also need to take into account the following points:

(1) Trees in smaller pots are liable to be shaded by larger containers, especially if the trees in the latter are stimulated to make more rapid growth.

(2) Small pots mixed with larger containers can easily fail to get enough water, and they may also dry out more rapidly. Thus care is needed to prevent additional water stress (C 41) occurring in some treatments.

(3) With complete randomisation, several replicates of one treatment may happen to be allocated to one small part of the area.

For problems (1) and (2), a compromise might be to randomise short *lines* of trees of the same treatment and pot size. Layouts such as *Randomised Blocks* (D 55 in Manual 4) can reduce problem (3), and *Latin Squares* can avoid it altogether.

What would the Blocks consist of?

In the pot size experiment, five triplets could be randomly assigned to each of 5 Blocks, making five sets of 15 plants. Each Block would be laid out together in a convenient and compact pattern within the whole experiment.

But supposing all the pots won't fit into one place?

If necessary, some of the Blocks could be put in one area, and the rest in a second place, according to the available space. As long as the whole Block is kept together, it would even be possible to put each Block in a separate place, as long as they were all reasonably similar.

Do experimental trees need special care?

Yes they do (C 48), in order to:

(a) avoid introducing extra, unnecessary sources of variation; ***and***

(b) lessen the risks of damage to individual trees, or of losing the whole experiment.

How about assessing and analysing the results?

In order to be able to judge properly (C 55, C 67-69), it is important:

(A) to **observe** frequently (and make brief notes about) what is going on; ***and***

(B) to **measure** or **count** some aspects of growth (C 67-68).

Some research workers **decide beforehand** what should be assessed, but others prefer to **choose at the time** what to measure and when to do it. If combined with regular observations, the second approach may allow one:

(1) to assess what looks particularly relevant to your experiment;

(2) to measure when differences between treatments are just becoming marked; ***and***

(3) to decide on priorities that are achievable with the people, equipment and time available.

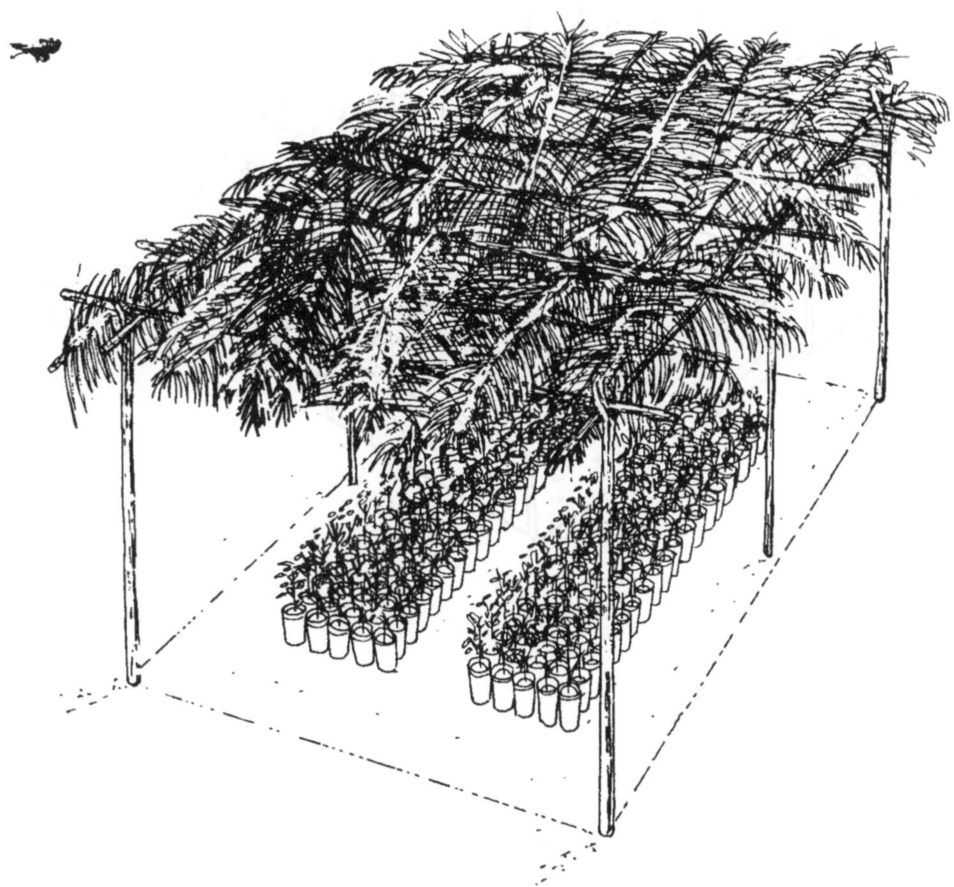

PLANNING A TREE NURSERY

Planning a tree nursery - *introduction: choosing the site* C 20

What is the most important feature of a tree nursery?

Producing environments in which young trees can thrive.

So does that mean that trees could be propagated anywhere?

Well, nursery sites do vary a great deal (C 2), but choosing a favourable spot is usually needed in order to:

(a) overcome problems and minimise losses during propagation (C 3, C 60);
(b) grow young trees with suitable root systems (C 4, C 11); ***and***
(c) produce enough good planting stock (C 34, C 47) at the right time for planting (Manual 5).

Does it matter whether the nursery is temporary or permanent?

No. The same general points apply (C 21-22), whether it is:

(1) near the house, growing small numbers of trees to plant on the farm;
(2) far from any buildings, producing enough trees to plant one area; ***or***
(3) on a main road, turning out hundreds of thousands of trees in a year.

How do I choose a good site?

Above all, the tree nursery needs to be near to an adequate supply of **water** (C 24), though not liable to **flooding;** and preferably also:

(a) not on a very steep **slope** (D 23 in Manual 4), nor completely flat land (D 12);
(b) on **soil** that is suitable (C 23), and near a supply of good topsoil (C 6);
(c) reasonably **sheltered** (C 25, C 46); ***and***
(d) **accessible** for supplies (C 24, C 51), and also to the expected planting site.

What is especially important about the water supplies?

(A) **Sufficient amounts reliably available,** even during dry weather when water use may be at its maximum;
(B) **Reasonably clean,** without having much:

(1) salt or toxic chemicals dissolved in it; ***or***
(2) fine soil particles, such as clay and silt (C 23; and D 12 in Manual 4), suspended in it; ***and***

(C) **Free or cheap,** because paying for mains water can be very expensive when large numbers of young trees are going to be watered frequently.

Supposing there is some salt in the water?

Traces of salt can sometimes be blown inland by strong winds, or left during exceptionally high tides, and it is best to avoid sites where this could happen.

Salty water from the sea or tidal estuaries is suitable only for trees like mangroves. However, these species need nurseries only if the natural stands have been largely removed (D 26 in Manual 4), or when other genetic origins or species are to be planted.

What if all the available land is steep?

Try to find a site which:

(1) is not on loose soil or where a stream might wash some away;
(2) is well protected by surrounding trees; ***and***
(3) can be reached easily.

Successful tree nurseries can still be produced that have a series of narrow terraced beds and standing areas for containers (D 65 in Manual 4).

Supposing the whole area is a flat plain?

The chief problems could be:

(a) poor drainage, so that the soil is sometimes waterlogged; ***and***

(b) the risk of more serious damage from flooding.

However, you might try:

(1) building up the height of seed beds and standing areas for potted plants;

(2) digging drainage channels (D 65 in Manual 4); ***and***

(3) making low walls to hold back flood water.

What other soils should I avoid if possible?

(a) **Very stony soils** (C 23), as a lot of effort would be needed to make them suitable.

(b) **Very heavy soils,** with a lot of clay in them, which would make nursery work much more difficult;

(c) **Extremely acid** or **alkaline soils,** with very low or with high pH (D 12 in Manual 4), which would need special treatment to grow all but a few tolerant species;

(d) **Highly degraded land,** requiring a lot of soil improvement for seed beds (D 22, D 32 in Manual 4); ***and***

(e) **Very exposed sites,** needing a lot of shade and shelter (C 25, C 41, C 46), while still remaining liable to storm damage (C 3).

How important is access for a nursery?

(A) By foot or by boat, it is essential;

(B) With a wheelbarrow or head pan, it is highly desirable;

(C) By motorised vehicle capable of travelling on rough terrain, it is very useful;

(D) If connected to a metalled road, it is a helpful bonus.

For access to the planting site, see C 21; and D 61 in Manual 4.

What supplies are needed?

Besides water (C 24), the main items generally include:

(1) supplies of topsoil, sand, old sawdust and other materials for making potting mixes (C 6, C 24); and sometimes soil and roots from older trees as *inoculum* (C 30-32);

(2) containers (C 6) and basic tools (C 51);

(3) poles, shading and fencing materials (C 41, C 46).

Building and roofing materials, basic furniture and extra tools may also be needed.

But supposing I can't get all those things?

If funds are limited, your nursery is very remote or the land is degraded, concentrate on:

(a) choosing the best site you can find;

(b) obtaining the most important items for good tree growth, but using methods that do not involve costly or bulky purchases; ***and***

(c) getting the young trees to the planting site in good condition.

What else is important about the site?

(1) Protection from climatic extremes (C 3, C 40-41);

(2) Convenience for carrying out regular watering (C 43) and checks (C 40); ***and***

(3) The size needed (C 22) for the number of young trees to be grown, allowing for the possibility of future expansion; ***and***

(4) The shape of the area chosen for the nursery in relation to planning an efficient internal layout (C 22).

Are there some other guide-lines?

(A) Before deciding where to put the tree nursery, and setting it up (C 26):

(1) look for advice in other nursery manuals (C 61-A), particularly any available for your locality;

(2) make a list of possible nursery sites, together with their advantages and disadvantages (C 54); ***and***

(3) discuss the options with other relevant people (C 53).

Planning a tree nursery - *temporary or permanent?* C 21

Isn't a permanent nursery always best?

No, it is often a good choice, but sometimes a temporary nursery makes more sense.

When would that be?

Provided that (given the seeds) people locally could raise good planting stock, temporary nurseries can be a good idea, for instance when:

(A) small numbers of young trees are to be grown near the house, perhaps under the shade of a short-duration crop such as bananas;

(B) plants will be wanted nearby for one or two seasons only; ***or***

(C) planting is in remote areas or difficult terrain where bringing in young trees would be very difficult.

What are the advantages of temporary nurseries?

(1) They can be set up near the planting site, so that the young trees:

(a) do not have to be carried far; ***and***

(b) can be moved just before the planting time (Manual 5); ***and so***

(c) may be subjected to considerably less stress (C 41, C 47).

(2) If the nursery is made by clearing a piece of woodland:
(a) the soil may remain relatively fertile during the period of use (C 23);
(b) trees may be left around the nursery, with perhaps a few scattered across it, to give protection from wind and sun (C 25).
(3) Establishing them is less dependent on the availability of substantial funds.

And what counts against temporary nurseries?

Particularly when they are remote, it may be harder to:
(a) provide the knowledge and training (C 50, C 52) needed for small, scattered tree nurseries to succeed, utilising the skills that have been learnt in another area;
(b) bring in the tools and materials needed (C 51); ***and***
(c) check regularly that the work is being done properly, whether the plants are growing well and when they will be ready for planting (C 40, C 47).

Could a temporary nursery be converted into a permanent one?

Yes, this might be possible, provided that it is appropriately sited, and:
(1) there is enough space available (C 22);
(2) the water supply is sufficient and reliable (C 24);
(3) access is adequate (C 20).

You could take this possibility into account when setting up a temporary nursery.

What are the advantages of permanent nurseries?

(A) Larger numbers of young trees can be grown, sometimes at a lower cost per plant;
(B) Planning and supervision of the work may be easier (C 40, C 50), reducing the risks of damage to the young trees (C 3, C 41);
(C) More tools and materials can be held (C 51), immediately available for use;
(D) It is easier to build up the experience and skills of a team of staff and workers (C 50, C 52), and to continue to benefit from the training received;
(E) Fences and buildings can be put up, and hedges and shade trees grown, which improve:
(1) the growing environments for the young trees (C 4, C 10-15, C 25);
(2) their protection from damage (C 3, C 25, C 46, C 60);
(3) the smooth day-to-day running of the nursery (C 54); ***and***
(F) Special facilities for research (C 15) or for valuable collections can be handled.

Do they have some disadvantages?

Yes. For instance, a permanent nursery:
(1) usually costs more to set up and run;
(2) may lose the fertility of its nursery soil (C 23) after some time; ***and***
(3) could run out of local supplies of components for making up good potting mixes (C 6, C 24).

If all the planting stock is grown in a single large nursery, an extra risk is that it might all be lost in a major incident, accident or attack by pests or disease (C 3, C 45).

So when should I choose a permanent nursery?

(a) If a suitable site is available not too far from the likely planting areas;
(b) When you expect an ongoing need for planting stock which temporary nurseries could not meet;
(c) If the nursery is to have a training function for the district;
(d) When research is to be carried out, or where careful supervision is needed for a particular tree species to be grown successfully; ***or***
(e) If a stockplant area is to be established nearby, to produce regular supplies of shoots to be rooted as cuttings (Manual 1).

Planning a tree nursery - *size, shape and layout* C 22

How do I decide on a suitable size for the tree nursery?

This depends firstly on how many young trees are going to be grown (C 63-A), and by what methods (C 2). For instance, when:

(A) **cuttings are to be rooted,** space would be needed for poly-propagators for rooting and for 'weaning' (Manual 1);

(B) **seeds are to be sown,** this might be done in:

(1) *seed trays* under cover;

(2) *seed beds* on the ground; ***or***

(3) directly in *containers*.

Except for (B 3), space will also be needed for growing on the young trees when they are potted up or transplanted (C 63-B).

Does the type of planting stock make a difference?

Yes, it does. The amount of space needed will vary quite a lot, depending on the set of species being grown, and whether the young trees will be planted as:

(a) container-grown plants;

(b) root-pruned soil blocks;

(c) striplings or stumps; ***or***

(d) bare-rooted plants.

Suppose I am planning a very small nursery?

If just a few container plants are to be grown in a temporary nursery, the size of its growing area can be simply determined by the space needed for each pot, multiplied by the number of pots.

What about a larger, permanent nursery?

For the growing areas themselves, there are six main points to consider:

(1) how many young trees are needed each year;

(2) the type of planting stock to be produced;

(3) what are the likely wastage levels from seeds and cuttings that fail to produce successful plants;

(4) roughly how long the trees will need to stay in the nursery before planting;

(5) whether their growth habits are slender or bushy, and so how much space each plant will take up; ***and***

(6) if planting can be done:

(a) only during one (or two) short periods of the year; ***or***

(b) for extended periods, or continuously.

Will space be needed for anything else?

Yes; it is also important to allow room (C 63-C) for:

(a) access paths amongst the beds of seedlings, transplants and potted plants;

(b) access roads, including turning, loading and parking places;

(c) buildings, sheds, a covered working place and storage areas for materials (C 24);

(d) hedges, shelter belts and shade trees (C 23, C 46).

What about future expansion?

Yes, this is an important point. Needs could include:

(A) growing unexpected extra batches of trees;
(B) putting up an additional building; ***or***
(C) being in a position to expand the nursery output later on, if required.

In any case, it is usually sensible to allow 25% more than the minimum calculated, so that one is not cramped for space to work in.

But isn't it a pity to make the nursery too big?

Yes, if:

(1) information is lacking on how many trees will really be needed;
(2) space is urgently needed for other permanent land-uses; ***or***
(3) costs of land are very high; ***but***

No, in most other circumstances, because it would be a pity if space had previously been available, but now was not, and yet:

(a) the nursery is proving successful and needs to expand; ***or***
(b) more flexibility is desirable in order to improve the internal layout.

How can I get a rough idea of the total area needed?

See sheet C 63-C for this.

Does the shape of the nursery matter much?

Not nearly as much as finding a suitable piece of ground (C 20). However, try and avoid:

(A) a long, narrow piece of land, which might mean a lot of extra carrying, and possible 'bottlenecks' that restrict working and access; ***and***
(B) a very wavy margin, which could give problems in laying out the beds, and lead to extra expense for fencing (C 25, C 46).

What is the best shape?

A piece of ground that:

(1) will allow east-west alignment of the beds, to allow more even light to the young trees, but is not more than three times as long as it is broad;
(2) has more or less straight margins; ***and***
(3) will allow entry at a convenient point in the internal layout.

Is the layout particularly important?

Yes, for all but the smallest nurseries, because a well-planned arrangement of the different parts helps to achieve a smooth 'flow' of materials into, work done in, and young trees going out of the nursery. So, having worked out how much space is needed for each part, you could:

(1) think about the ***order*** in which the various jobs will be done;
(2) make some rough sketches of possible layouts that would permit a good 'flow';
(3) walk the ground, make some rough measurements, and see whether some fit in better with the terrain, points of access and existing shade and shelter; ***and***
(4) choose the one that looks likely to:
  (a) provide enough working space and avoid blocks to easy moving of materials;
  (b) minimise unnecessary walking and carrying (C 20).

So a good layout can save time?

Yes it can. For example, good planning will save you nearly an hour if a job takes:

only one minute less for each plant, and there are 50 of them; ***or***
just 5 seconds less per plant, and there are 700 plants.

With jobs that are done frequently (C 54), the time saved by an efficient nursery layout can amount to a substantial amount each week.

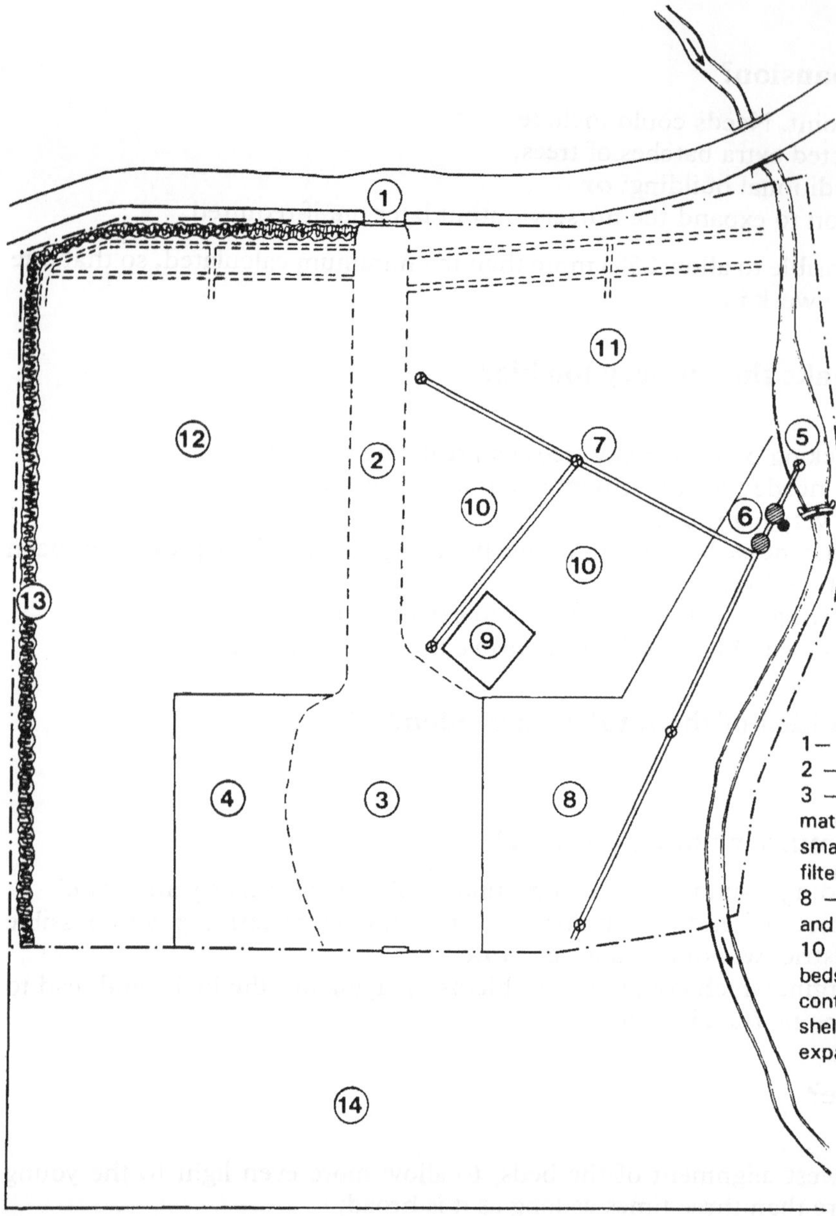

1— double gate and fence.
2 — nursery road, with paths off it.
3 — turning area. 4 — storage of materials. 5 — water source and small dam. 6 — water pump and filter. 7 — pipes and taps.
8 — covered working area, store and office. 9 — shade house.
10 — seed beds. 11 — transplant beds. 12 — standing area for container plants. 13 — hedge for shelter. 14 — space for later expansion.

Are there any other advantages of planning the internal layout?

Yes, there can be several, for example:

(A) the growing areas could be better protected (C 25);
(B) parts with better soil or drainage (C 23) could be used for beds rather than for roads or buildings; ***and***
(C) less damage to the young trees is likely to occur if there is room to walk between them, and if any taps for hoses are carefully positioned.

Is it easier to get it right if the nursery is new?

Yes in some ways, for there may be the opportunity to:

(a) learn from visiting other nurseries; ***and***
(b) do some planning before choosing the site, nursery size or internal layout.

However, in a new nursery there may be less experience amongst staff, and fewer skilled workers (C 53). One possibility would be to:

(1) start off in a small way with temporary, movable structures;
(2) modify the layout as experience is gained; ***and***
(3) build a more permanent facility after a year or two of experience.

How should the actual laying out of the nursery be done?

See sheet C 26.

Planning a tree nursery - *nursery soils* C 23

Can a tree nursery be made on any soil?

One might be able to manage where it is unsuitable, but it is much better if you can choose a site with a favourable soil.

Supposing all the plants will be grown in seed trays and containers?

Provided you can make up good soil mixes (C 6), then you could still achieve satisfactory tree growth and suitable root systems for planting even with a poor nursery soil. Indeed, access to a plentiful supply of components for making up good seed and potting mixes (C 24) can be more important than the type of nursery soil. However, the smooth running of the nursery (C 50-54) can be hampered if the ground:

(A) gets very sticky after rain, or is liable to flooding;

(B) blows about and covers everything with dust in the dry season; ***or***

(C) is so hard or rocky that it is difficult to make the beds and paths, or to put in poles to support shading.

So a good nursery soil is preferable?

Yes it is. In particular, it keeps open the option of using nursery beds of various kinds (C 22, C 26).

What are the features of a good nursery soil?

The main points are to have:

(1) a favourable soil *texture,* depending especially on the proportions in it of soil particles of different sizes;

(2) sufficient organic matter and nutrients, and an appropriate acidity; ***and***

(3) adequate depth, aeration and drainage.

It can also be an advantage if the nursery soil already contains micro-organisms which form close associations with tree roots that are beneficial or essential for good tree growth (C 30-32).

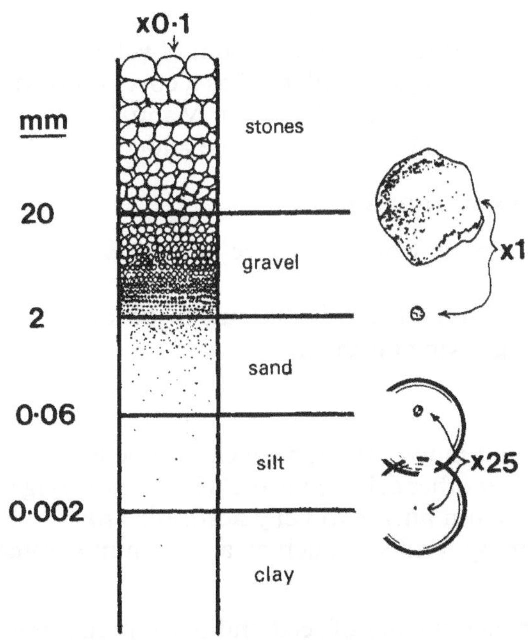

Which kinds of soil particles make the texture unsuitable?

If possible the nursery soil should not:

(a) have a lot of large stones in it;
(b) contain a lot of clay; ***or***
(c) consist mainly of a single size of particle.

A light, moderately sandy loam is often most suitable (D 12 in Manual 4).

But what if I can't find a place with a favourable soil texture?

If the only available sites are on:

(A) ***soils rich in silt or clay,*** you could:

(1) dig some coarse sand into the topsoil of seed and transplant beds;
(2) put a thin layer of grit or fine gravel on the soil surface after sowing seeds, to discourage damping-off disease (C 45); ***and***
(3) cover paths with a thick layer of coarse gravel.

(B) ***soils that are very sandy,*** one could:

(1) dig organic matter into the topsoil of the beds, such as composted plant and animal wastes or sieved 'black soil' (C 33); ***and perhaps also***
(2) add some fine silt.

(C) ***stony sites,*** you might:

(1) dig stones out of the topsoil; ***and***
(2) use them for foundations for buildings, laying concrete, making drains or putting in the bottom of poly-propagators (A 31 in Manual 1).

Wouldn't this be too much work?

Not if it makes it possible for the tree nursery to produce good planting stock! Anyway, most of the work is not on the entire nursery, but just the top 10-15 cm of the growing areas.

Will such improvements also add to soil fertility?

Yes, because improving the soil texture generally means that:

(a) earthworms and other important decomposers (D 10, D 13 in Manual 4) are more likely to be recycling nutrients and improving soil texture;
(b) the roots of the young trees can grow, branch and collect water and nutrients better (C 11, C 13-14); ***and***
(c) there is less likely to be a need to add fertilisers (C 33).

How else could I have a fertile nursery soil?

(1) The most fertile sites tend to be those on volcanic soil, but good sites can often be found where richer soil has been deposited near the bottom of valleys (D 12 in Manual 4).
(2) Nitrogen-fixing species (C 32) can be planted as a cover crop when beds are not in use, as hedges (C 46) or as shade trees (C 41; and D 41 in Manual 4).
(3) Green manure or appropriate fertilisers could be added (C 33).

Does the acidity of the soil matter?

Somewhat acid soils are generally best for a tree nursery, with a pH between about 4.5 and 6.5 (D 12).
Very acid, or **alkaline** soils may make it difficult to produce good trees, except for local species that are adapted to growing on them.

Can one alter the pH?

To a limited extent one could modify the topsoil of seed beds, by working in some:

(a) crushed limestone or other calcareous rock, lime, basic slag or acid-reducing fertilisers such as sodium or calcium nitrate to **very acid** soils; ***or***
(b) urea or acid-forming fertilisers such as ammonium sulphate or nitrate to **alkaline** soils.

It is somewhat easier to change the pH of seed and potting mixtures (C 6).

How deep should a nursery soil be?

The deeper the better in most circumstances, with a minimum of 8-10 cm, except where all the young trees are in raised beds or containers.

And what about soil aeration and drainage?

Uncompacted, freely-drained soils, with plenty of air spaces in them, are most suitable for nurseries, because:

(1) the root systems of the young trees are unlikely to become waterlogged (C 11) after watering or during rainy weather; ***and***

(2) the paths and access roads are less likely to be cut up if they have to be used soon after rain.

Soils with *impeded* drainage will be less suitable on both these counts, while excessively freely-drained soils may dry out very rapidly.

Type of soil:	**Sandy gravel**	**Sandy loam**	**Silty clay**
Texture:	loose	good	hard when dry, sticky when wet
Air supply to roots:	excessive	plentiful	restricted
Drainage:	excessive	free	impeded
Water retention:	poor	good	very good, but soil cracks when dry

How can drainage be improved?

(A) Choose a slightly sloping site for the nursery (C 20), and align the paths and beds downhill.
(B) If necessary, put in some drains. These could be:

(1) **open ditches,** preferably with only a slight gradient, and with barriers at intervals (D 65 in Manual 4) and perhaps a layer of stones on the bottom, to reduce erosion and also retain useful water;

(2) **'blind' drains,** (trenches filled back in with stones, between which the water can run), which could form some of the nursery paths; ***or***

(3) **plastic pipes with perforations,** which can be buried beneath the beds or paths.

(C) Add some fine gravel, grit or coarse sand to the nursery beds, and consider raising them above the level of the paths. Put plenty of gravel on these.

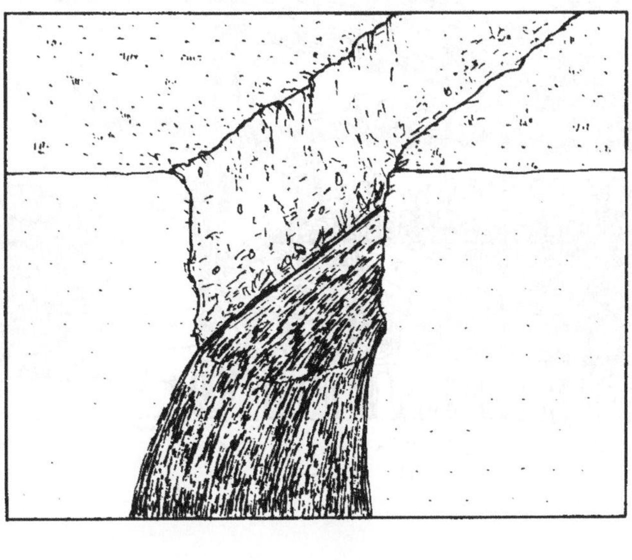

Open drain.

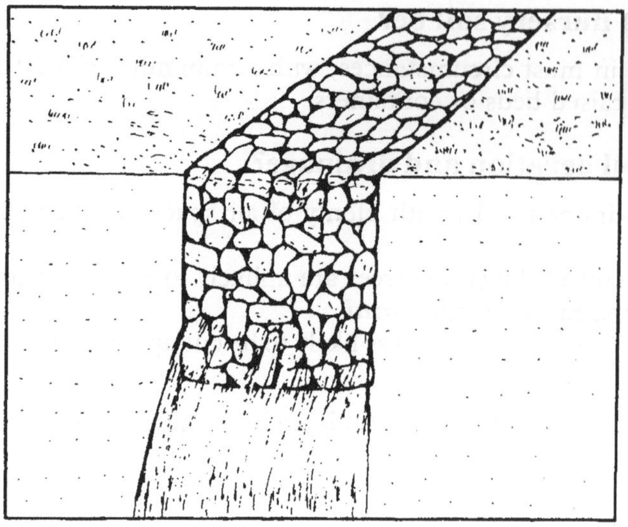

Blind drain.

What can I do if drainage is excessive?

(a) Align the beds and paths along the contour;
(b) Plant hedges to trap soil and slow run-off;
(c) Add organic matter or some silt to the beds.

On a steep site, you could consider making a terraced nursery (C 20; and D 65 in Manual 4).

Is there anything else to consider?

Try and avoid sites that are:

(A) very windy and without existing trees (C 25);
(B) close to the high tide level, in case salt is washed or blown in during exceptional weather (C 20);
(C) near to a place where toxic wastes have been or are being dumped; ***or***
(D) containing persistent weeds, or near a source of their seeds (C 44).

Planning a tree nursery C 24
- *supplies of water and soil mixes*

Isn't water supply the most important thing about the site?

Yes, that's right. There are very few places where one can rely on rainfall always watering the trees in the nursery adequately. Having a nursery water supply is essential for:

(A) rooting cuttings in poly-propagators or other systems (Manual 1);
(B) germinating small seedlings in a protected environment (Manual 2); ***and***
(C) watering the young trees during rainless weather, so that they survive and grow well (C 13, C 43).

Water is also needed when preparing potting soils (C 6) and for cleaning tools (C 45, C 51).

What about a small, temporary nursery near a house?

This could be supplied mainly from waste water from the home or a school, provided that:

(a) there will be enough of it every day;
(b) it does not contain very large amounts of detergent or soap, or smaller amounts of substances harmful to young trees; ***and***
(c) clean water is used for poly-propagators, seeds and germinating seedlings.

How much water will be needed in a larger tree nursery?

Young trees in containers may need watering at intervals from twice a day to twice a week, depending on the species, the container size, the potting mix and the weather. Each seed tray or pot might take between about 10 millilitres (ml) and a litre each time, and there is bound to be some waste.

Young trees in beds also need watering, usually a little less frequently. You could estimate the quantity at 10-20 litres per m^2 of growing area.

So a nursery with 20 m^2 of beds and 1000 young potted trees might sometimes require as much as a thousand litres a day.

Can I do anything to reduce waste of water?

Yes, several things can be done, including:

(1) storing water when plenty is available for times when it is scarce and demand is greatest;
(2) channelling rainwater to help keep the deeper levels of soil moist, or using polythene sheeting to direct excess water into a collection pond;
(3) using sunken transplant beds (provided that they will not become waterlogged):
(4) avoiding over-watering the young trees (C 11, C 42);
(5) choosing less wasteful ways of watering; ***and***
(6) training people to avoid pouring water on paths, and to look out for and mend leaks promptly (C 50, C 52).

Which kinds of water are suitable?

(a) A stream or river;
(b) A freshwater lake;
(c) A pond or small reservoir created by building a dam;
(d) Rainwater, collected from the roofs of buildings or with polythene sheeting, and stored in large barrels or tanks;
(e) A well or borehole; ***or***
(f) Mains water, though this can be very expensive (C 20).

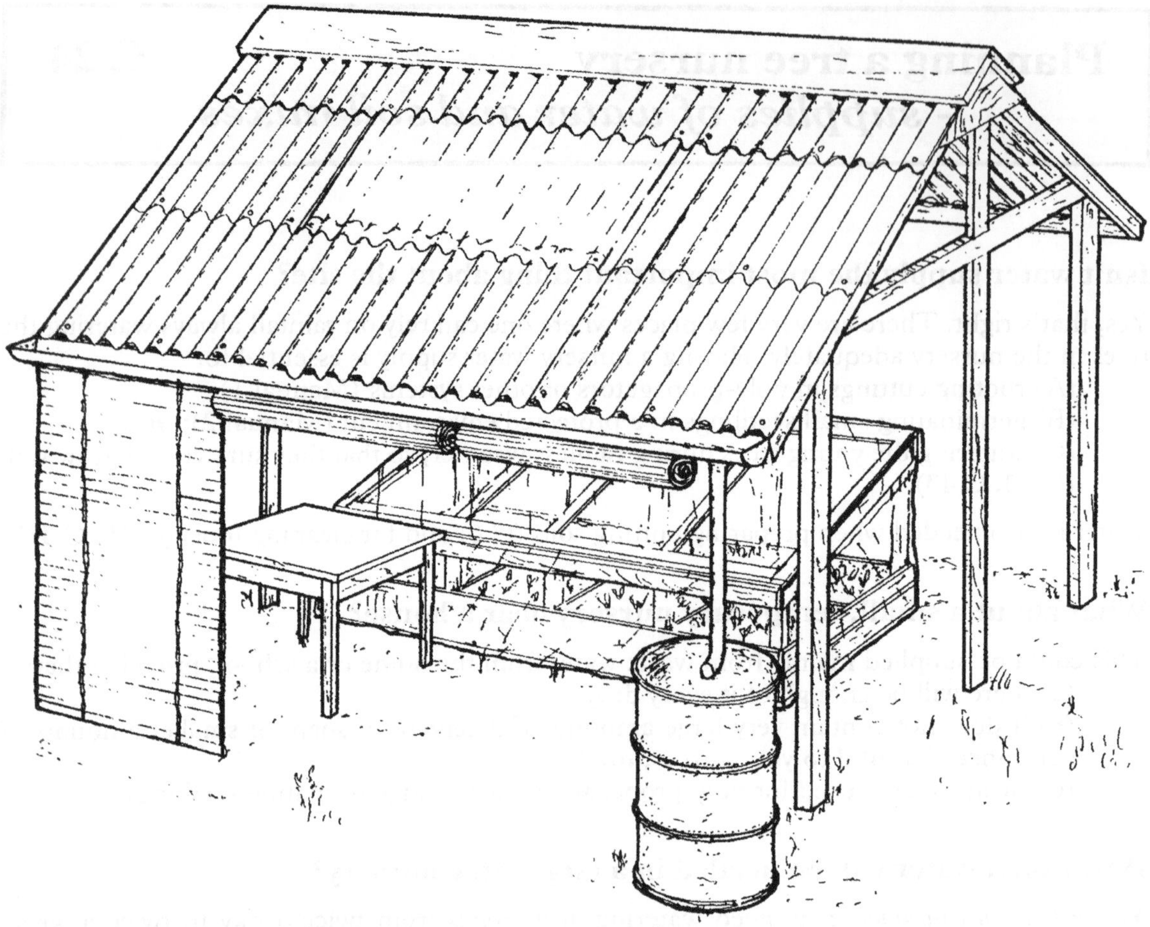

What are the key points about the water supply?

(A) **Sufficient quantity:** enough water is needed to be able to:

(1) water every potted plant twice a day;
(2) water all the seed trays, seed beds and transplant beds once a day;
(3) store some water as a reserve;
(4) allow for the washing of people and vehicles; ***and***
(5) build in a safety margin (for example of 50% extra).

(B) **Reliability:** great problems involving emergency supplies will result if the water runs low or stops completely during very dry weather, and the young trees might still be damaged or even die.

(C) **Reasonable cleanness:** the water needs to be free of:

(1) a lot of silt, clay, dust or other fine particles suspended in it;
(2) dissolved pollutants, salt, oil or other chemicals that are toxic to most plants.

Can't some of these things be cleaned out?

(1) **Suspended particles:** most of these can be fairly easily removed by:

(a) arranging for the water to be taken from a sheltered part of the water supply;
(b) leaving the water to stand before using it; ***or***
(c) passing it through a simple filter (for example, gravel followed by coarse sand) and the regularly and frequently cleaning this out.

(2) **Dissolved substances:** although extracting them is usually possible, it is often too expensive for a tree nursery. It would be better to try and arrange things so that less of these enter the water supply.

However, the supply does not need to be up to the standard of pure drinking water.
Note: staff and workers should be clearly informed about this (C 52).

Couldn't standing water give some problems?

Yes, you might find a build-up of:

(a) **mosquito larvae,** unless suitable fish are present, or a few drops of paraffin (kerosene) are added to water containers;

(b) **algae** or an excess of other small micro-organisms in the water, which can usually be checked if water containers can be covered;

(c) **water weeds,** which can be removed periodically with a hooked stick or a net;

(d) **dead leaves,** which can be excluded or removed if there is a tendency for the water to become foul-smelling.

Should the water supply be inside the nursery?

For all nurseries, it is a big advantage if the site is chosen so that the water supply is **within or close to the nursery,** provided that it will not cause flooding.
For small nurseries, you could have some buckets handy to catch waste water.
For large nurseries, if the water supply is below the nursery, think about pumping water to a high point, where it could flow by gravity to supply a series of taps throughout the nursery.
On some dry or steep sites, it might be necessary to carry or pipe in the water from a distance.

Hydraulic rams do not require an external energy source, since they use water pressure to pump a little water to a higher level.

And it matters how the watering is done?

Yes, watering is a skilled job, though it can be done in various ways (C 43).

What supplies of materials are needed for making soil mixes?

A variety of components, described in sheet C 6, that when mixed together provide:

(a) a suitable mixture of particle sizes for the roots to branch into (C 4, C 11);

(b) good drainage and aeration with adequate water retention (C 23);

(c) enough organic matter and nutrients.

How important are soil mixtures?

They are usually vital to the growing of good planting stock. Their most important uses are to:

(1) mix up good potting soils for trees grown in containers (C 6);

(2) make a suitable germination medium for plants raised in seed trays (Manual 2);

(3) prepare a good rooting medium for cuttings (A 35 in Manual 1); ***and***

(4) improve the condition of the nursery topsoil in beds of various kinds (C 23).

How much of each will be needed?

Surprisingly large amounts. For instance, about 1 m^3 of a component forming only 25% of a potting mix would be needed to pot up 1000 cylindrical pots, 14 cm wide and 18 cm tall (C 63-C).

Where should I put all these components?

Since you could be handling a lot of heavy materials, it is well worth:

(a) choosing a place where they would be most convenient for use (C 20, C 22);

(b) building a set of simple wooden or cement block partitions to keep each component separate;

(c) getting in a large supply of those commonly used; ***and***

(d) planning ahead (C 54) to continue getting reliable supplies.

What other things might I need?

Other materials, tools, record sheets and notebooks (C 51, C 54, C 64-67).

Planning a tree nursery C 25
- *protecting the whole nursery*

What does the nursery need protection from?

At an early stage in planning, think about what can be done to minimise the risks of:
- (A) flooding;
- (B) strong winds;
- (C) fire;
- (D) browsing animals; ***and***
- (E) thieves and vandals.

How can I protect against flooding?

In several ways, including:
- (1) careful choice of the nursery site (C 20), avoiding:
  - (a) very flat, low-lying or swampy land;
  - (b) a place very close to a river that might rise to that height; ***or***
  - (c) areas, near to the coast or tidal estuaries, which might be flooded when an unusually high tide occurs; ***or if necessary***
- (2) building a protective wall of mud, bricks or concrete blocks; ***and***
- (3) keeping a watch when the flood risk is high.

What kinds of damage can flooding cause?

(a) Driving out the air from the spaces between the soil particles, depriving the root systems of oxygen (C 11, C 23); ***and***
(b) Depositing clay, silt or sand all over the nursery; ***and perhaps also***

(c) Bringing salt into the nursery soil (C 24); ***and***
(d) Breaking or washing away of the young trees, if the current is strong.

Can anything be done about strong winds?

Powerful storms are always likely to do some damage, but the extent can often be reduced, and damage from lesser winds more or less eliminated, by:
- (1) choosing terrain (C 20; and D 12 in Manual 4) that provides natural shelter;
- (2) retaining trees near the nursery that will check the force of the wind;
- (3) planting shelterbelts around and hedges within the nursery (C 46); ***and***
- (4) if possible, growing tree species that are less liable to damage from wind.

Added protection can be provided by putting up a covered area for working under while it is raining (C 22), and for germinating very small seedlings (Manual 2), building poly-propagators for rooting cuttings (Manual 1), and putting up shade houses or greenhouses for research plants (C 15, C 48).

Which species would be most suitable?

(A) Those that will act as a good windbreak or give light shade, and not produce heavy branches that might break off;
(B) Some of the species to be grown in the nursery, which might later act as a handy source of inoculum for mycorrhizas or root nodules (C 30-32).

What might be most at risk from fire?

The young nursery trees and any buildings or vehicles, especially:
- (a) in dry forest, savanna and grassland areas;
- (b) if planted trees nearby are of species that assist the spread of fire;
- (c) when fire protection lines have not been maintained;
- (d) where the nursery is temporary (C 21), or when no-one is around for most of the time;
- (e) when local communities are unaware of or opposed to the tree-planting.

For fire protection, see C 3; and also D 11, D 16 and D 66 in Manual 4.

***Note:* Staff may also be at risk in certain situations!**

How about animals?

The three most important ways of protecting against browsing animals are:
- (1) putting up fences (or even walls), or planting thick, spiny hedges (C 46);
- (2) tethering or guarding domesticated animals; ***and***
- (3) exchanging information about the need to protect young trees, and the kinds of domesticated and wild animals that pose the greatest risk (C 53).

What might thieves want to steal?

Items liable to be stolen might include:
- (A) vehicles, equipment and tools;
- (B) containers, fertilisers and other useful materials; ***and***
- (C) young trees.

Occasionally vandals can cause mindless damage without stealing anything.

Is there anything that can be done?

You might consider the need for:
- (a) always putting away tools and smaller items in a locked shed;
- (b) keeping vehicles and larger equipment in a locked garage or near someone's house;
- (c) employing a watchman at night, and over weekend and holiday periods;
- (d) trying to involve the local community in the tree-planting project (D 5 in Manual 4); ***or***
- (e) building a high fence with lockable gates (C 46).

Are there other things that might harm my young trees?

Yes, they could be damaged for various reasons (C 3, C 60), for example by:
- (1) sudden exposure to bright sun or drying winds (C 13, C 41-42, C 47);
- (2) poor watering or unsuitable regimes (C 43);
- (3) competition from weeds (C 44);
- (4) attack by diseases and pests (C 45);
- (5) contamination of the water or soil supplies (C 24) with toxic substances; ***or***
- (6) other kinds of accidents (see D 66 in Manual 4).

These risks can be greatly reduced by regular checking (C 40).

Planning a tree nursery - *setting it up* C 26

What do I need to do before actually setting up a tree nursery?

(A) If possible, visit local nurseries and talk to the people running them (C 53);
(B) Read sheets C 2, C 4, C 6 and C 20-25, and other information about nurseries (C 61);
(C) Estimate the approximate numbers of young trees you will be growing, and the areas needed (C 63 A-C);
(D) Look at several possible sites and finally choose the best (C 20);
(E) Make the necessary arrangements for using, leasing or owning the land;
(F) Think about how to have adequate supplies of water and components for making up soil mixes (C 24, C 63-D), and sources of seeds and shoots for cuttings (C 5); ***and***
(G) Consider who is going to do the work (C 52), and where any money will be found (C 2).

How should the work start?

The first thing to do is to choose the most effective internal layout for your nursery (C 22).

And what comes next?

(1) Felling unsuitable trees and clearing unwanted existing vegetation, while keeping other trees and shrubs that could provide useful shade (C 41), protection (C 25) and temporary cover crops over parts of the nursery that will not be used at once (C 22);
(2) Doing any major earth movements for roads, drainage or terracing that may be needed (C 20, C 23);
(3) Arranging the water supply (C 24), including any ponds, water tanks, pumps or piping;
(4) Planting hedges and additional trees for shelter, shade or protection, and fencing the nursery if this will be needed (C 25, C 46).

Should I now start marking out the different parts of the nursery?

Yes. This could be done by:

(a) pacing out on the ground to see if the planned layout (C 22) still fits, or whether it needs modifying;
(b) measuring accurately, and marking out the different parts with pegs or poles;
(c) making minor adjustments as necessary.

What different parts could the nursery have?

In a small, temporary nursery (C 21-22), there might only be growing areas and paths.
In a larger, permanent nursery, there could be:

(1) the growing areas themselves;
(2) paths, drains, hedges and shade trees;
(3) a covered working area, storage shed, office and washroom;
(4) storage space for stocks of materials; ***and***
(5) an access road, together with turning, loading and parking space.

Which sort of growing areas could there be?

(A) Shaded poly-propagators for rooting and weaning cuttings (Manual 1);
(B) Shaded seed beds and a covered area for germinating seeds (Manual 2);
(C) Transplant beds with removable shading for growing root-pruned soil blocks, striplings, stumps or bare-rooted planting stock (C 22);
(D) Standing ground for containers, also with removable shade; ***and perhaps***
(E) Stockplant areas, for producing plenty of suitable shoots for vegetative propagation;
(F) A shade house or greenhouse to provide special environments for research (C 48).

The remainder of the nursery area consists of spare ground in case of extra demand, and for later expansion (C 22).

How big should the beds and standing ground be?

The **width** should generally be between about 0.8 and 1.2 m, so that weeding, watering, checking and moving plants can be easily done from the two sides;
The **length** should not exceed 15 m, or access may become difficult.

See sheet C 63-B for calculating the total areas needed.

What about setting up the growing areas?

Preparation will vary according to the types of growing area chosen. For instance:

(A) *Poly-propagators* need to be in a sheltered place where they can be checked frequently (C 40; and A 31 in Manual 1);

(B) *Seed beds* might need to be raised 10-20 cm above the level of the paths to give good drainage. They could be started with gravel, continued with a suitable ordinary soil mixed with coarse sand and topped off with a fine seed germination mixture (Manual 2). Covered germination areas could be part of the sheltered working area, or in any convenient sheltered spot;

(C) *Transplant beds* could be made similarly, but topped off with a thicker layer of a less fine mixture. Alternatively, the nursery topsoil could be used as it is, if suitable (C 23). Sunken beds are sometimes used in dry areas, to conserve moisture (C 24), but be careful that the plants do not become waterlogged;

(D) *Standing ground* for pots might be put on a part with poorer soil;

(E) *Stockplant areas* are best as lines, far enough apart to allow harvesting of cuttings, pruning and weeding, and perhaps interplanted with a soil-improving shrub (A 32);

(F) *A shade house or a greenhouse* needs to be carefully planned, with adequate drainage and good shading, and with some tables or staging for smaller plants (C 48).
For studies of effects of the environment on tree growth (C 15), the ability to control ventilation and perhaps soil temperature might be useful, while for trees with chronic insect problems (C 45) the house may need to be constructed with mosquito netting.

How about setting up the shading?

(a) Decide whether to put up plastic shadecloth over part of the nursery, such as the seed and cutting propagation section and the working area, or to use leaves or mats (C 41);
(b) In the second case, choose between high or low shading; ***and***
(c) Allow for the supporting poles when calculating the width of paths and beds.

What else do I need to think about before growing the first plants?

(1) Confirming that the water supply is reliable (C 24);
(2) Getting in a supply of materials and enough tools (C 24, C 51);
(3) Starting to build a nursery team by beginning to train any staff and workers in how to carry out the various jobs reliably and well (C 52);
(4) Making up a seed germination mixture (C 6) and rooting medium (A 35 in Manual 1);
(5) Obtaining suitable seed (C 5) and arranging where to collect the first shoots for cuttings;
(6) Beginning the keeping of records (C 54) about:
 (a) where the young trees originated (C 64);
 (b) when particular batches of plants were started (C 65); ***and***
 (c) daily and weekly propagation checks (C 50, C 66).

Which species should I grow first?

It is a good idea to try out the beds and the running of the nursery with some well-known soil-improving crop plants first, such as beans or groundnuts. This is likely to:

(A) increase soil fertility, while giving useful produce to the nursery team (C 52);
(B) show up any problems with the growing environments or techniques (C 3); ***and***
(C) act as a training session before starting to grow the trees (C 50).

MICRO-ORGANISMS, NUTRIENTS AND TREE GROWTH

Micro-organisms, nutrients and tree growth C 30
- *introduction: close associations with roots*

How do micro-organisms influence trees?

All trees are affected by micro-organisms. Some of them, for instance:

(1) are decomposers, helping to break down litter and release nutrients (C 14; and D 13 in Manual 4);
(2) live immediately outside the fine roots; ***or***
(3) grow on the surface of leaves.

What are 'close associations with roots'?

Other kinds of micro-organisms in the soil have much closer relationships with trees, entering the roots and changing their structure.

Are they common?

Yes, close associations with woody plants appear to be widespread throughout the world. The great majority of tropical trees are thought to have at least one such association.

How important are they?

They are vital, especially on poor soils (C 62-C,D). Most of the micro-organisms closely associated with tropical trees play a particularly important part in:

(a) the nutrition of the individual tree (C 14), and so how well it grows;
(b) quantities and cycling of nutrients in the ecosystem (D 13, D 32 in Manual 4);
and probably
(c) improving the uptake of water (C 11, C 13).

NOTE: Many **soil-improving trees and shrubs,** which can reclaim degraded sites (D 22 in Manual 4), have at least two types of special associations with their roots.

What kinds of micro-organisms form close associations?

(A) Fungi;
(B) Bacteria; ***and***
(C) A few other groups, such as the *actinomycetes.*

But aren't these the things that cause disease?

A few kinds of fungi and bacteria can cause disease (C 45);
Most of them just break down litter, and are not harmful; ***and***
Some species form these special associations, with direct benefits for themselves and their host trees.

What do the useful micro-organisms gain?

(1) A ready supply of sugars and other organic substances for energy and growth;
(2) A protected place where they can live; ***and***
(3) Extra opportunities of spreading and multiplying.

But won't that *reduce* the growth of my trees?

Not necessarily. It is often shortage of nutrients that limits the growth of trees in the tropics, though many other factors may also be involved (D 10-16 in Manual 4). Except where young trees are heavily shaded or unhealthy, plenty of sugars can be available for this 'trade-off' with the fungus in exchange for nutrients.

The general principles are becoming clearer about close associations with roots (C 62), but so far only a limited amount of research has been done on specific details.

Are there names for these associations?

Mycorrhizas are close associations between fungi and tree roots (C 31);
Root nodules are close associations between bacteria and tree roots (C 32); ***and***
Actinorhizas are close associations between actinomycetes and tree roots.

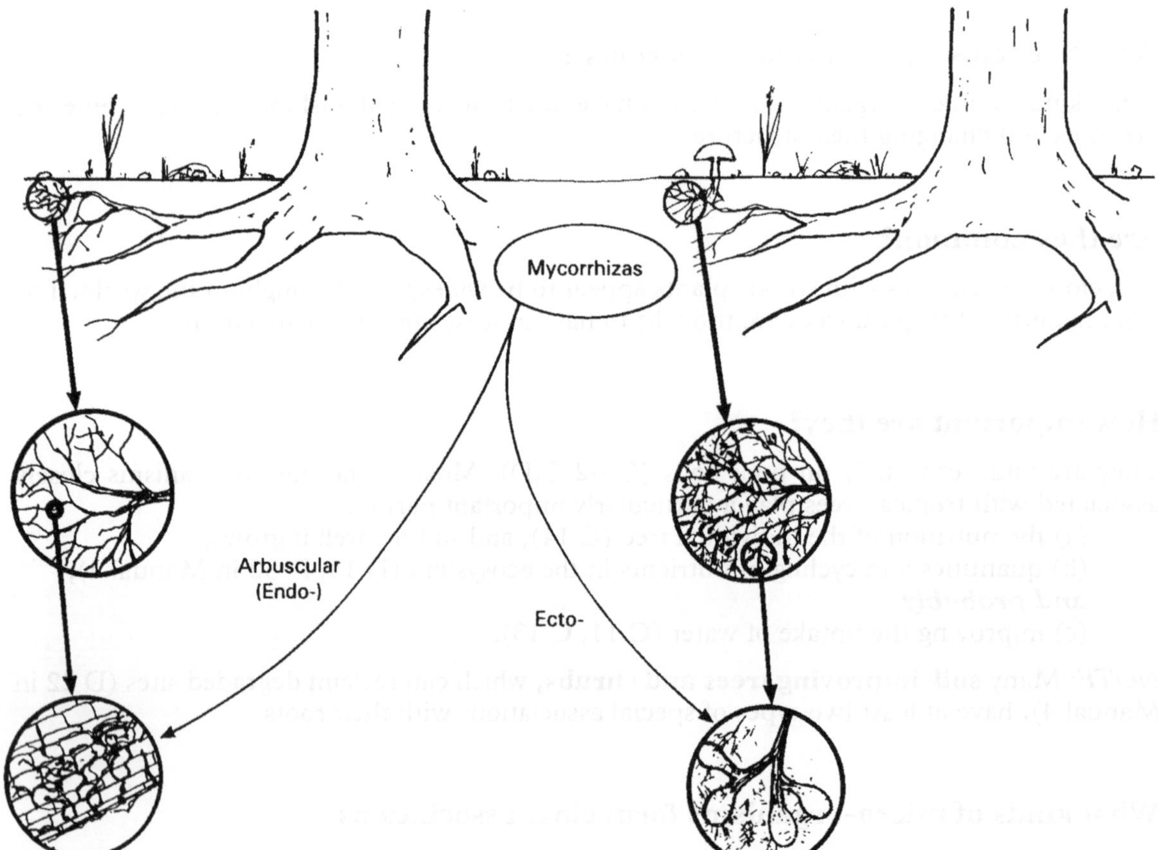

What are mycorrhizae like?

Mycorrhizal roots contain fungal tissue and have an altered structure. There are two different kinds:

(a) **Ectomycorrhizas,** in which fungal threads form a sheath *outside* fine rootlets of the tree, and penetrate *between* the root cells.
(b) **Arbuscular mycorrhizas** (endomycorrhizas, vesicular-arbuscular mycorrhizas or VAMs), where the fungi grow *inside* the cells of the roots. They occur in the outer part of roots that have not yet thickened and become woody, though not in the conducting cells within (C 10-11).

The fungi forming both types of mycorrhiza also produce:

(1) a very extensive system of very fine threads *(hyphae)* running and branching through the soil; ***and***
(2) very large numbers of resting spores.

Can you see ectomycorrhizas quite easily?

Yes, they can usually be spotted with the naked eye or a hand lens. Ectomycorrhizal rootlets are often short and relatively thick with blunt ends.

Most of the fungi involved form large fruiting bodies ('mushrooms') that are easily seen at certain times of year, though connection to a tree is difficult to establish.

Is it harder to see the arbuscular mycorrhizas?

Yes it is, for they cannot be seen with a hand lens, nor do the fungi involved form large fruiting bodies. However, infected roots are very common, and can be detected by using a simple staining technique and then looking at samples under a microscope.

How do mycorrhizas benefit the tree?

Through nutrients and probably water being passed to it from the fungus.

But surely a tiny fungus can't help the growth of a big tree!

Surprisingly it can, because:

(a) its threads can be very extensive, providing a **large surface area** for substantial amounts of nutrients and water to be taken up from the soil;

(b) some ectomycorrhizal fungi can break down, absorb and pass on nutrients from the soil that otherwise are **unavailable to tree roots.**

In fact, these fungi may be so important that some trees cannot thrive without them (C 31, C 60), and mycorrhizas can even form a substantial part of the root systems of some tropical trees. Equally, many of the fungi cannot thrive without the trees.

Which nutrients are passed to tropical trees by mycorrhizas?

Many kinds of nutrients are probably transferred, but **phosphorus** is especially important. This is one of the three elements needed in large amounts by trees (C 14), but in tropical soils it is often scarce or locked away in an unavailable form.

How can mycorrhizal fungi utilise 'unavailable' nutrients?

(1) The extensive network of fungal threads may **reach** immobile substances, such as phosphorus, more efficiently than a tree root *(one estimate suggests they may reach 400 times more soil); **and***

(2) Some ectomycorrhizal fungi produce special enzymes (C 14) that can **break down** resistant organic substances in the litter or soil, which other decomposers cannot do.

What about root nodules?

These are modified, roundish bodies on the roots, a few mm in diameter, in which large numbers of bacteria occur. Such nodules are easily seen and are well known on the roots of leguminous crop plants like beans and groundnuts. They also occur on many (though not all) leguminous trees, particularly those in the families Mimosaceae and Papilionaceae (C 32, C 62-D).

Which nutrient are they mainly involved with?

Nitrogen - the nutrient required in largest quantities by plants (C 14), but often readily lost and not freely available in tropical soils.

How do the bacteria help?

Nitrogen gas forms about 80% of the atmosphere, but in this form it is unavailable to plants. However, bacteria belonging to the group called **rhizobia** are able to turn nitrogen gas into soluble nutrients. These *nitrogen-fixing* micro-organisms can therefore add to the **total amount** of nitrogen that is available.

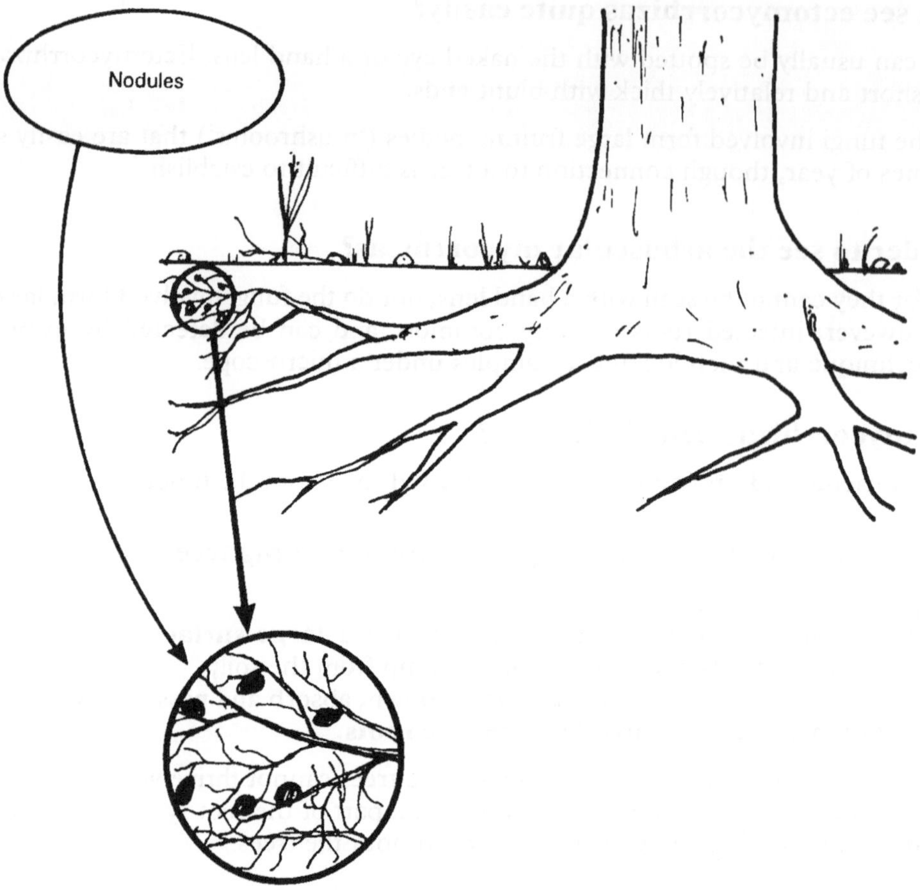

Can trees get hold of this extra nitrogen easily?

Yes, by soluble nitrogen passing:

(A) directly from the nodule into the rest of the tree; ***and***

(B) indirectly when nodules decay, so that the nutrients are released into the soil by decomposers and can then be absorbed by the tree's roots.

So you get a free fertiliser!

Yes, that's right. This is one of several reasons for planting mixtures of different species (D 30, D 53 in Manual 4). If nitrogen-fixers are included, the total amount of nitrogen that is cycling in the ecosystem is increased, which means that other trees and crops without root nodules may grow better, especially on poor soils.

Is that especially true of soil-improvers?

Most of these species, besides having nitrogen-fixing root nodules and mycorrhizas, are quick-growing colonisers that can become established on open or degraded land (D 22, D 32 in Manual 4).

Some examples with confirmed nitrogen-fixing ability are associations with *Calliandra calothyrsus, Gliricidia sepium, Inga jinicuil, Mimosa scabrella, Leucaena leucocephala, Sesbania grandiflora* and species of *Acacia, Albizia and Erythrina.*

What about actinorhizas?

These occur as root nodules on the roots of soil-improving trees such as *Alnus, Casuarina, Allocasuarina* and *Coriaria.* The micro-organisms belong to a kind called *Frankia,* and like rhizobia they help the tree to acquire nitrogen.

Research has shown that effective formation of nodules can vary considerably from one provenance of a tree species to another (C 5).

Micro-organisms, nutrients and tree growth C 31
- *mycorrhizas: fungi plus tree roots*

What have mycorrhizas got to do with tree nurseries?

A good nursery root system is one of the keys to growing and planting trees successfully (C 4, C 11, C 34). If the root systems of the planting stock are already mycorrhizal, the young trees are likely to establish better.

Which kinds of tropical trees produce ectomycorrhizas?

(a) All **dipterocarps,** for example *Dipterocarpus, Hopea* and *Shorea.*
(b) Some **leguminous trees,** like *Afzelia, Brachystegia, Eperua, Gilbertiodendron, Intsia, Microberlinia* and *Tetraberlinia* in the family Caesalpiniaceae; and *Aldinia, Pericopsis* and *Swartzia* in the Papilionaceae;
(c) Many **conifers,** such as pines and species of *Araucaria;*
(d) A few other **broadleaved** trees, including some eucalypts and oaks.

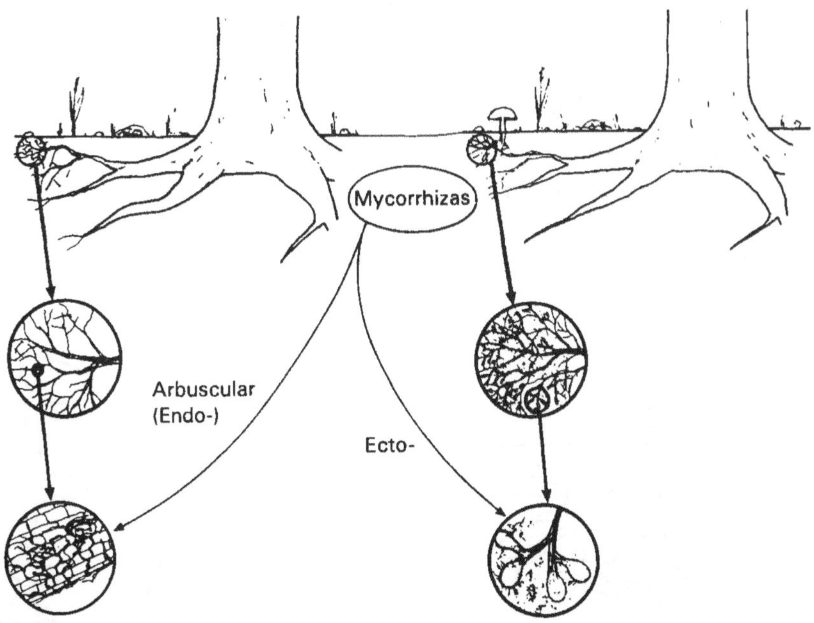

And which trees form arbuscular mycorrhizas?

Around 95% of all tropical forest trees are thought to do so. A few species of *Acacia, Afzelia, Eucalyptus, Hopea* and *Intsia* can even form both types of mycorrhiza.

Are there some groups of trees that don't form mycorrhizas?

It is thought that trees in the families Lecythidaceae, Proteaceae and Sapotaceae do not do so.

Does it matter which fungus forms the mycorrhiza?

It often seems to make quite a lot of difference, because particular species of fungi appear to be associated with certain groups of tree species. For example, ectomycorrhizas formed by the fungus *Pisolithus tinctorius* and the conifer *Pinus caribaea* have been studied by several workers (see Lapeyrie and Högberg, 1994 in C 62-C).

In ectomycorrhizas, different groups of fungi appear to be involved as the tree gets older.

Is there much variation within one kind of fungus?

Yes, there can be. Some *strains* of a fungus may form **more effective mycorrhizas** than others, just like some *clones* of trees make better growth (A 11 in Manual 1).

There is some evidence that a specific strain of a fungus might combine particularly well with a particular tree clone.

Can I tell which kinds of fungi are present?

Mats of threads can sometimes be seen amongst wet litter or in the topsoil, but these could belong to non-mycorrhizal fungi, and are too fragile to trace far.

Identifying the fungal species in the soil or the tree roots is possible, but it requires detailed microscopic study of roots, threads and spores, and sometimes the culture of individual fungi in the laboratory.

So it's not much use to me!

Yes, it could be.

Why is that?

Although identifying species may be difficult, it is relatively straightforward to:

(a) detect the **presence or absence** of mycorrhizas; ***and***

(b) estimate the **frequency of infection** of roots with mycorrhizas.

But how can I make use of that?

(A) By knowing whether your young trees are infected; ***and***

(B) By inoculating them with mycorrhizal fungi, which:

(1) is not difficult; ***and***

(2) might make a substantial difference to their performance.

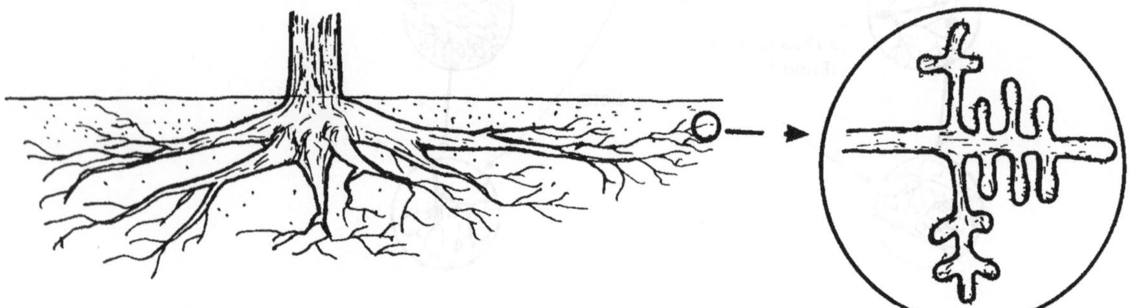

Ectomycorrhizas

How is infection with ectomycorrhizas detected and assessed?

By collecting samples of fine roots, and:

(a) finding out what percentage show the characteristic groups of short, slightly swollen ends to fine roots (C 30); ***and***

(b) checking under the microscope for the presence of the network of fungal threads covering the rootlets.

And what about arbuscular mycorrhizas?

Infection is checked by collecting samples of fine roots, treating them with a *clearing agent* that makes it possible to look through them, and using a coloured *stain* to show up the fungal cells. The proportion of roots containing fungus can then be estimated under a microscope.

What does one use to inoculate the root systems of trees?

There are four possible sources of mycorrhizal fungi, which use an *'inoculum'* from:

(1) topsoil containing fine roots, collected under trees of the particular species that are growing in natural forest, savanna or a well-established plantation;

(2) nursery or container soil that has recently been used to grow that tree species;

(3) spores or pieces of the fruiting bodies producing them, or *cultures* of kinds of fungi that have been shown to make effective ectomycorrhizas with the tree species.

(4) *soil cultures* of an arbuscular mycorrhizal fungus. For instance, strains of *Glomus* are beginning to become available.

Which source is the best?

Source (1) is a useful starting point if thriving stands of the tree species are available;
Source (2) is convenient if there is an existing nursery producing good plants;
Source (3), where available, is likely to prove the best way of obtaining ectomycorrhizas that can improve tree establishment, reliably and substantially.
Source (4) - simple methods have been developed for multiplying the quantities of inoculum for arbuscular mycorrhizas. These consist of growing crop plants such as cowpea, millet, sorghum or maize for a few months in a sterilised (C 44-45) and inoculated potting mix in large containers. These enriched soil cultures can then be added to soil mixes as required.

Note: national regulations and international certificates govern the import of soil and roots from one country to another, because of the dangers of transferring pests and diseases.

Inoculating soil with spores from the fruiting bodies of ectomycorrhizal fungi.

Isn't it hard to inoculate the young trees?

No. One just mixes a small amount of the inoculum into the topsoil of seed-beds (C 23) or with the soil mix used to fill seed trays or pots (C 6, C 42). Failing this, it is possible to put a little inoculum into the hole when a tree is planted in the field (Manual 5).

Which method is likely to be best?

Ideally, one would inoculate the germination or potting soil with **a mixture of known strains** of fungi. Then the young seedlings or rooted cuttings should form particularly effective mycorrhizas early in life, before they are exposed to the extra stresses of hardening (C 47), transport, planting, competition and browsing.

How would I know that inoculation had done anything?

By doing a small experiment (C 7, C 15) in which you potted up comparable plants in the same soil mix, with and without a fungal inoculation of about 5% by volume.
Both sets would need labelling (C 54) and looking after carefully (C 40, C 48).
For the layout, you could keep the two treatments in similar conditions near to each other, though not so close that spores could easily be transferred by water splashing on to the soil. Similarly, the containers could be raised clear of the ground to prevent seepage of water from pot to pot.

If the results were positive, you might do a more detailed experiment to compare several sources.

How big a difference can mycorrhizas make?

For ectomycorrhizas, introductions of pines, cypresses and other species into new areas have sometimes failed (C 60) without a first inoculation of nursery or forest soil from the country of origin, followed by transfer of infected soil from nursery to nursery. Increased growth of inoculated pines has also been found in several trials during the first 1-2 years after planting.

For arbuscular mycorrhizas, a suitable local fungus may more often be available, though inoculation can often still be beneficial. For example, in establishment trials with *Terminalia prunioides* in Kenya more than twice the number of plants given a mixed inoculation in the nursery survived through the first long dry season than in the uninoculated controls.

What other trees have been studied?

Among the kinds of trees that have responded to arbuscular mycorrhizal inoculation are:

(1) *Cedrela, Gmelina, Khaya* and *Tabebuia;*
(2) many leguminous trees, including *Acacia, Albizia, Calliandra, Cassia, Gliricidia, Leucaena, Pithecellobium, Prosopis, Sesbania* and *Tamarindus;*
(3) various fruit trees, including avocado, papaya and mango; ***and***
(4) shrubs such as cassava *(Manihot esculenta).*

Arbuscular mycorrhizal inoculation is standard nursery practice for large *Citrus* plantations in some countries, but more research will be needed before this stage can be reached with other tropical tree species.

When is inoculation especially needed?

It is likely to be most important when:

(a) **introducing species** from a different part of the same country or region, or from another part of the world *(especially for ectomycorrhizas);*
(b) the soil is naturally **poor in nutrients** *(such as infertile sandy areas);*
(c) planting is being done to **restore degraded land** *(D 22, D 32 in Manual 4);*
(d) the planting site has been recently **disturbed,** for example by frequent cutting, burning or farming; or near mines, buildings and roads *(especially important when most or all of the original trees have gone, and there are a lot of weeds);* ***and/or***
(e) site preparation is to be done with **heavy machinery** *(particularly if the surface soil is likely to be compacted, broken up or buried).*

Why is this so?

In all these cases the right kinds of mycorrhizal fungi may be absent or in short supply. Their growth and spread are usually interrupted by the removal of many trees, which exposes and disturbs the soil. Other fungi, associated for instance with persistent weeds, may have become common.

Planting introduced colonisers (D 14, D 32 in Manual 4) could help local species to regenerate naturally (D 2 in Manual 4), by restoring numbers of arbuscular mycorrhizal fungi.

When would inoculation be less important?

(A) For fertile planting sites, where mycorrhizas may play a less prominent role;
(B) When the nursery has already been producing thriving young trees that are mycorrhizal;
(C) If wildings (C 2) are collected with some of the local soil, especially directly under the seed tree. It has been found that naturally regenerating seedlings can even be sustained in dark, dry and nutrient-poor conditions through ectomycorrhizal 'bridges' with the parent tree.

Collecting wildings and soil.

How long can I keep the inoculum?

It is generally best to collect fresh material each time. However:

(1) fruiting bodies of ectomycorrhizal fungi can be stored dry for a year, as can spores extracted from them; ***and***

(2) nursery soil infected with arbuscular mycorrhizas can be multiplied repeatedly to provide inoculum.

What should I do if I run into problems with mycorrhizas?

You might contact (C 53):

(A) growers nearby to see whether they have solved a similar problem;

(B) the nearest forestry or agriculture department for advice; ***or***

(C) a local non-governmental organisation for help in finding out whether the problem has been studied. (See also C 62-C; and D 71 in Manual 4.)

Do any nursery procedures affect mycorrhizas?

(a) **Soil sterilisation:** sterilising the nursery soil generally kills all the beneficial mycorrhizal fungi and decomposers as well as any disease-causing agents, insect pests and weed seeds (C 44-45). So if the soil should need to be sterilised (by heating or by chemicals), inoculation with mycorrhizal fungi might well be needed afterwards.
(b) **Control of plant diseases:** special care is needed with *fungicides,* or the mycorrhizal partner will be damaged or eliminated. However, they are not all equally toxic to fungi. For instance, it has been found with citrus that *thiazole* should be avoided.
(c) **Insecticides:** avoid excessive spraying, and try and prevent the chemical dripping down on to the soil. If any *systemic* insecticides are to be applied as a soil drench, you might do an informal trial first, and perhaps an experiment to compare different techniques and concentrations (C 15).
(d) **Fertilisers:** use sparingly or not at all (C 33), or you may reduce the likelihood of successful mycorrhizas forming, or of those that do being effective.

Restoring degraded land with a pure stand of a soil-improving tree, *Acacia mangium*.

Micro-organisms, nutrients and tree growth C 32
- *nitrogen-fixing root nodules*

What are nitrogen-fixing nodules?

They are modifications to some of the fine roots of a tree, in which closely associated micro-organisms live. Generally these are bacteria called rhizobia, or sometimes actinomycetes called *Frankia.* Both of them have the ability to change atmospheric nitrogen into available nutrient form (C 30).

Aren't root nodules found mainly in annual crop plants?

They have been widely studied in legumes such as groundnuts and soybean, and in temperate zone grazing and crop plants. However, there are 18,000 or so leguminous species, most of which are trees, many of them tropical species.

Which kinds of trees have nodules?

(1) **Leguminous trees:** current research suggests that 90% or more of the hundreds of genera of tropical trees in the Mimosaceae and Papilionaceae may be found to have root nodules that fix nitrogen. However, less than a quarter of trees in the Caesalpiniaceae may do so, though in *Chamaecrista,* for instance, effective nodulation is the rule.
(2) **Non-leguminous trees:** a few genera are known, including *Casuarina* and *Parasponia.*

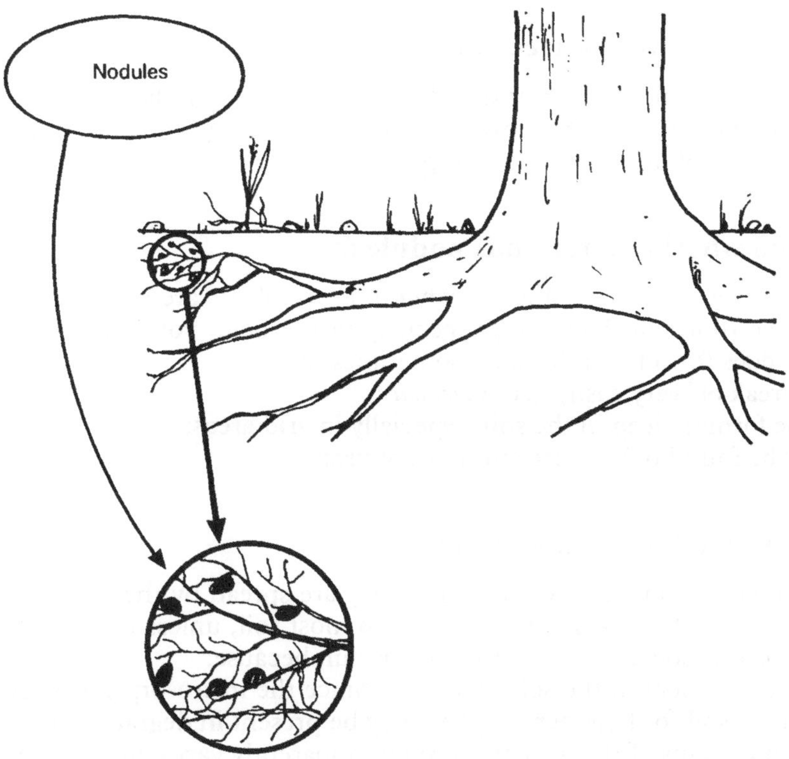

Are those used in agroforestry nitrogen-fixers?

Many of them. For instance Egli and Kalinganire (1988) (C 61-D) list the following for Rwanda: *Acacia albida, A. sieberiana, Albizia chinensis, A. lebbek, Calliandra calothyrsus, Leucaena leucocephala, Pithecellobium dulce* and *Prosopis chilensis* in the Mimosaceae; *Cajanus cajan, Erythrina abyssinica, Gliricidia sepium, Sesbania sesban* and *Tephrosia vogelii* in the Papilionaceae; with *Alnus acuminata* and *A. nepalensis* nodulating with *Frankia.*

Which legume trees *don't* form nodules?

Amongst those so far studied, the following are thought ***not*** to do so:

Caesalpiniaceae: *Bauhinia, Caesalpinia* and some species of *Senna (Cassia);*
Mimosaceae: *Adenanthera,* and perhaps certain species of *Acacia* and *Parkia;*
Papilionaceae: *Vatairea, Vataireopsis* and Brazilian *Pterocarpus.*

So are leguminous trees specially important?

Yes, they are. The many kinds that form effective nodules play unique roles through:

(A) adding to the **total amount of nitrogen** available within most tropical ecosystems (C 30; and D 10, D 13 in Manual 4);
(B) acting as **soil-improvers** in farmland and for reclaiming degraded soil (D 21-22 in Manual 4); ***and***
(C) improving the **general soil fertility** in nurseries (C 23) when used as hedges (C 46), shade trees (C 41) or windbreaks (C 25; and D 41 in Manual 4).

Does the planting stock itself need root nodules?

Having nodules on the young leguminous trees is likely to make a considerable difference if:

(1) they are effective in fixing nitrogen;
(2) the potting mixes (C 6) and nursery beds (C 23) are not very rich or heavily fertilised (C 33); ***and***
(3) the trees are being well looked after (C 40).

Research suggests that both the nitrogen content and the growth rates of nursery trees can be increased by root nodules.

How about when the young trees are planted out?

Already having effective nodules should add to the likelihood of planting stock surviving and establishing well. Such trees might grow well even in poor sites, and could soon contribute to the growth and yields of other trees and crops.

Is it easy to tell whether a tree has nodules?

Often they can be seen without difficulty when nursery plants are potted up, or by checking the outside of the root ball. However, especially in the field, you could miss nodules even though the tree does form them, because sometimes they:

(a) may break off very easily (as in *Acacia*);
(b) can be formed deep in the soil, especially in arid areas;
(c) might be found only at certain times of year.

Do young trees need to be inoculated?

Like other bacteria, rhizobia are so small that they are spread by the smallest current of air. Some of them are therefore likely to be present in most soil, unless it has just been sterilised (C 44-45). However, inoculation can often be useful, because:

(A) numbers present in the soil vary a lot. Since the micro-organisms generally do not survive in the soil for long periods, few may be present in degraded sites, for instance;
(B) different strains of rhizobia and *Frankia* apparently vary considerably in their ability to form effective nodules; ***and***
(C) as with mycorrhizas (C 31), there may not be inoculum in the local soil that will prove effective with introduced trees and shrubs.

For instance, inoculating with rhizobia has been recommended for ***Albizia lebbek, Pithecellobium*** *dulce* and *Prosopis chilensis.*
Adding certain strains of *Frankia* to potting mixes in the nursery trebled the wood production of *Casuarina* trees in the field (See Reddell *et al.,* 1989 in C 62-D).

What is the best source of inoculum?

You could try:

(a) adding soil (and any root nodules you can find) from a thriving plot of leguminous trees to your seed beds and potting soil;

(b) if possible, using strains of micro-organisms already developed for tropical crops. These are likely to be more effective than those for the temperate zone; ***or***

(c) if available, using strains that are recommended for the particular tree species (already developed for *Leucaena* and certain species of *Acacia*).

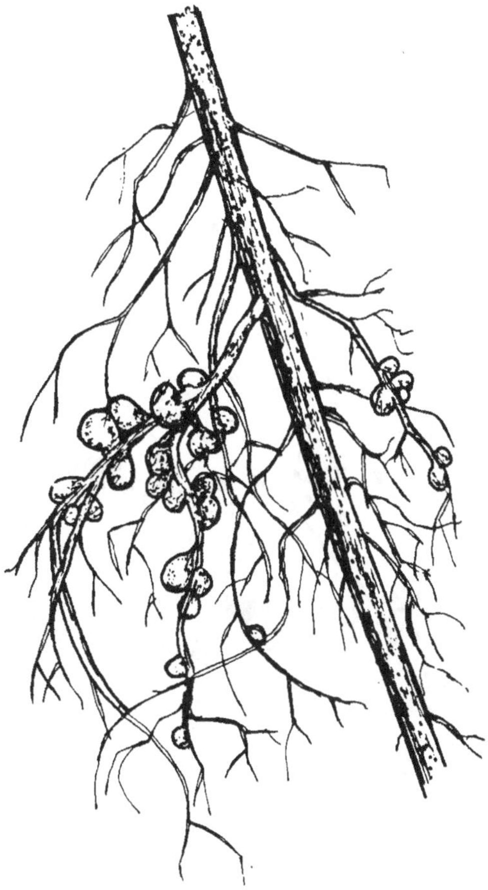

Root nodules.

How does one do the inoculation?

Either by:

(A) thoroughly mixing around 5% of soil and chopped roots with nodules in the seed or potting mix; ***or***

(B) dusting seeds with a very small amount of powdered inoculum.

What about doing an experiment?

A pot plant experiment (C 15) could be very useful, comparing treatments such as:

(0) control - just the ordinary potting mixture;

(1) the same potting mixture + an inoculum of root nodules from groundnuts mixed into it;

(2) the same potting mixture + an inoculum of soil from under thriving plot of the desired tree species;

(3) the same potting mixture + an inoculum of soil from under a different, local leguminous tree species.

You might use 15-20 pots for each treatment, giving a total of 60-80 pots. Keep these under similar growing conditions, but slightly apart from each other and raised off the ground to reduce the risk of spreading bacteria from one treatment to another. Avoid fertilisers or a very rich potting mix.

Do trees vary much within a species in nodule formation?

Yes, this seems to be likely. Considerable variation has been recorded *between* different provenances (C 5; and Manual 2); and also *within* a provenance. So, if clonal cuttings are available (A 11 in Manual 1), these might be the best experimental plants.

Can trees have root nodules and mycorrhizas?

Yes, some of them do (C 30); for example *Acacia nilotica* and *Leucaena leucocephala.* There is evidence that, without the mycorrhizas:

(a) the root nodules may not develop well; ***and***

(b) nitrogen fixation may be restricted.

It is likely that the phosphorus collected by the mycorrhizas allows normal production and functioning of the nodules.

A (above) — bare soil in degraded farmland.
B (below) — soil restored 18 months after direct sowing of *Leucaena*.

Micro-organisms, nutrients and tree growth C 33
- *wastes, composts and fertilisers*

Why do wastes have anything to do with tree nurseries?

Some wastes can be useful in various ways, for example by providing:

(1) containers of various kinds;

(2) suitable components for soil mixes (C 6) and rooting media (A 35 in Manual 1) that are free or cheap;

(3) valuable inputs of nutrients and/or organic matter to improve the fertility and structure of the nursery soil (C 23);

(4) mulch for seed and transplant beds to reduce soil temperatures and drying out;

(5) shade and shelter for young trees (C 41).

Waste water and rainfall can also be collected or encouraged to remain in the nursery soil (C 24) rather than quickly disappearing by evaporation or run-off.

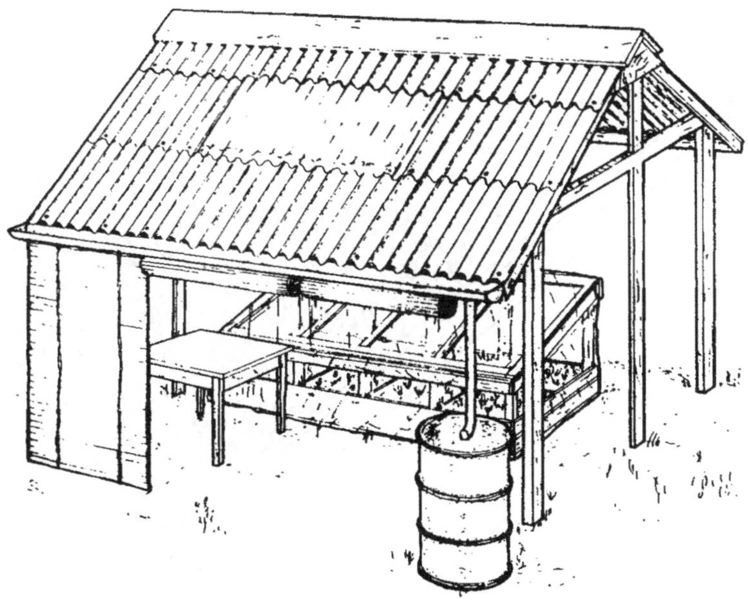

What sort of old containers could be useful?

For instance:

(a) cleaned oil-drums for storing water (C 24);

(b) plastic containers, cut and given holes so that they can be used as plant pots (C 6); ***and***

(c) used fertiliser or other large polythene bags, for purposes like storing soil mixes or carrying plants.

Which waste materials can go into soil mixes and rooting media?

(A) **Town wastes** that have rotted into 'black soil' can provide organic matter and nutrients;
(B) **Sawdust** that has weathered for about a year and **rice hulls** add organic matter and improve the soil structure, though they contain fewer nutrients;
(C) **Fibre** from coconuts and from the inside of rotting palm stems are useful, richer components; ***and***
(D) **Composts,** prepared in the nursery out of most kinds of plant and animal waste, can be the most valuable of all additions.

What can be used to improve the nursery soil?

As well as the wastes just mentioned, you might improve the soil in transplant beds by:

(1) digging in leaflets from branches of leguminous trees (C 32) when finishing the beds;

(2) growing a green manure or a nitrogen-fixing crop in the bed before putting in trees (C 26); ***or***

(3) putting on a mulch when a particular bed is not in use.

Fresh sawdust might be used to build up a wet, low-lying part of the nursery (C 23).

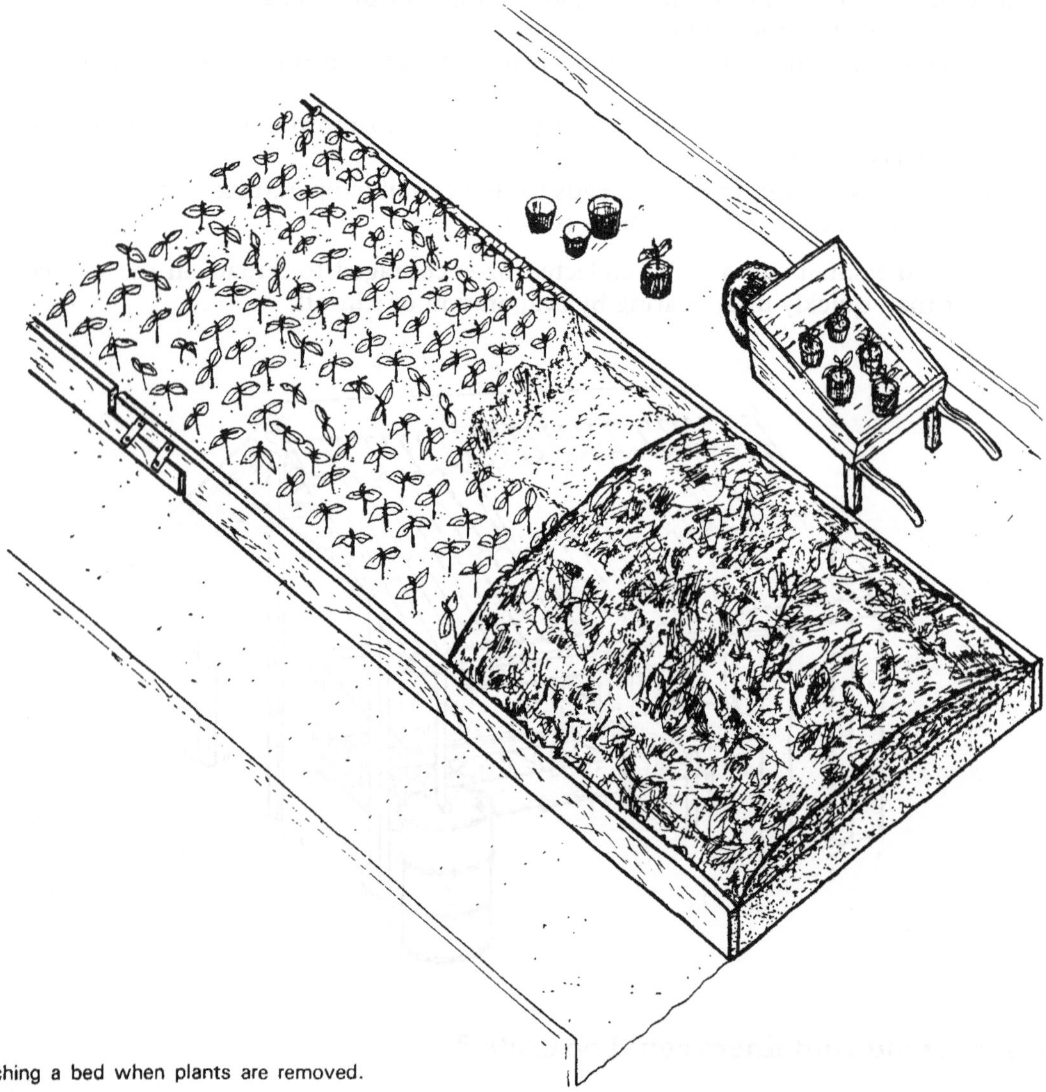

Mulching a bed when plants are removed.

How should a mulch be made?

Various kinds of soil covering can be used (see Goor and Barney, 1968 in C 61-A; and Manual 5), including cut grass, other kinds of leaves, sawdust, bark chippings or pieces of black polythene sheet.

Which wastes are useful for shade and shelter?

Cut poles and palm fronds, banana leaves or tall grasses could be useful to put up low or high shade over young trees (C 41).

Pieces of strong clear or white polythene sheeting from the covers of furniture and bedding are often suitable for building poly-propagators (A 31 in Manual 1), or protecting delicate seedlings from rain and wind (C 41).

Might some wastes be harmful?

Yes, they could be. Here are some examples to avoid:

(a) materials that could pollute the nursery soil (C 23) or the water supply (C 24);

(b) broken glass, thorns or spines, which can cause cuts and infections if they get into the potting mix or the nursery topsoil;

(c) weeds that have ripening fruits or persistent underground organs (C 44);

(d) plants and soil that are contaminated with spores carrying diseases, or with the eggs or resting stages of insect pests (C 45); ***or***

(e) water that contains a lot of nutrients and is full of bacteria or algae.

Is there anything that can be done about these things?

(1) Some sharp items can be sieved out (C 51) from components such as 'black soil', and disposed of carefully;

(2) Soil and plant material contaminated with pollutants, weeds, pests or diseases could be sterilised (C 44-45), burnt or taken far away from the nursery; ***and***

(3) Water can be cleaned by running it through gravel or sand.

What are composts?

They are made by mixing together plant and perhaps animal wastes to rot, in a pile, a large container or a pit (say 1.5 m deep).

The heat given off by the micro-organisms which break down these organic wastes can make the compost hot enough to kill some weed seeds, spores and insect eggs.

How does one make a good compost?

It is important to:

(A) mix several different kinds of material together;

(B) keep the compost heap moist by watering it in dry weather; ***and***

(C) turn it over from time to time, especially if it becomes very wet.

More details can be found in GTZ (1976) and Napier and Robbins (1989) in sheet C 61-A).

When is a compost particularly valuable?

Being rich in nutrients and organic matter, well-made compost is an especially useful addition when:

(a) the nursery or potting soil is hard, very sandy or full of clay (C 23);

(b) it is naturally infertile, or has been degraded (D 22 in Manual 4);

(c) there are enough people at the nursery to collect the materials to make it; ***and***

(d) alternative materials are unavailable, too distant or too expensive.

For any nursery, compost is a simple way of replacing some of the continual outflow of nutrients in planting stock.

Does a mulch do that as well?

Yes, it is one of the reasons for using it. Besides adding organic matter and nutrients, mulching could also help in the nursery by:

(1) **suppressing weed growth** in:

(a) seed beds before and after sowing, and transplant beds before and after use;

(b) containers in which young trees will stay for some time;

(c) planted stockplant lines that are being repeatedly harvested for cuttings (A 27 in Manual 1);

(2) **protecting seed beds** containing fairly large seeds against:

(a) overheating and rapid drying out of the surface soil;

(b) the direct impact of heavy raindrops on germinating seedlings;

(c) washing and/or blowing away of fine soil particles; ***and***

(3) **reducing the deposition of silt** on nursery paths and roads.

Mulch is also very useful when young trees have just been planted out (Manual 5).

Compost heap.

Wouldn't mulching encourage young seedlings to rot?

It might do if:

(a) it was applied during very wet weather;
(b) large, thick leaves were used; ***and***
(c) the mulch contained a lot of spores of the *damping-off* fungus (C 45).

But it would be unlikely to when:

(1) used in drier weather;
(2) lighter, chopped material was used; ***and***
(3) the mulch was lifted to aerate it after heavy showers.

Wouldn't it be simpler just to add fertilisers?

Yes, this can be an easy way of adding nutrients; but most fertilisers:

(A) do not provide the other benefits of composts and mulching; ***and***
(B) depend on external inputs, rather than sustainable management by recycling existing nutrients.

But surely fertilisers might be needed to get good growth!

Yes, sometimes they certainly could be. However, it is important to realise that fertilisers can have several disadvantages, such as:

(a) costing a lot to buy, and being heavy to transport to the nursery;
(b) causing damage to the roots of the young trees if too much is applied;
(c) often being quickly washed away during watering and by heavy rainfall; ***and***
(d) ending up contaminating water supplies (C 24).

What would be good reasons to add fertilisers?

They could be added to:

(1) **growing nursery plants,** if there are problems because of nutrient shortage or a need to change the pH (C 6, C 23, C 60);

(2) **potting mixtures** (C 6), especially when it is known that the:

(a) components all lack a particular nutrient (C 14), or more than one of them, so that tree growth would be seriously hampered;

(b) young trees need to grow in the same pots for a considerable time; ***or***

(c) rates of growth need to be increased, so that trees will be large enough at planting time (C 34);

(3) **stockplants** that are being repeatedly cut back (A 27 in Manual 1); ***and***

(4) **research trees,** to maintain good, uniform growth (C 7) or as treatments in experiments (C 15).

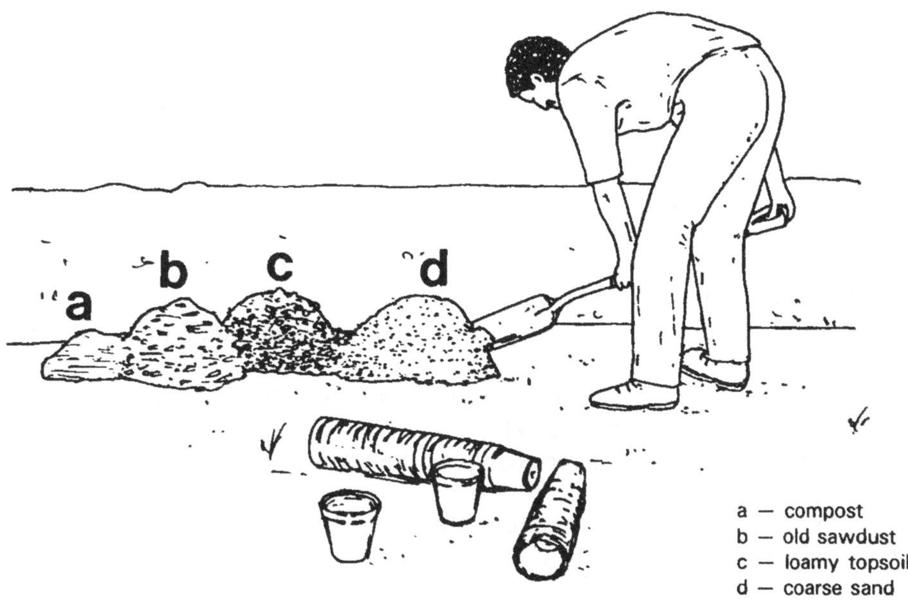

Which ways can they be applied?

(A) **As ordinary fertilisers** in the potting mix or scattered lightly over a bed;

(B) **As 'slow-release' pellets** in the soil, which last longer;

(C) **Dissolved in water,** which is often useful, because:

(1) it can be done during the regular watering (C 43), for example every 1-4 weeks;

(2) nutrients can be applied at times when rapid growth is desired; ***and***

(3) 'feeding' with nutrients can be stopped or changed when the trees are to be hardened (C 47); ***or***

(D) **As a foliar feed** of very dilute micro-nutrients (C 14), sprayed on the foliage.

Whichever method is used, the fertilisers should be balanced (C 14) and not too rich in nitrogen (C 34).

How much fertiliser should I add?

This depends a great deal on what is already being supplied to the young trees in the nursery soil or potting mix, in the water supply and by closely associated micro-organisms (C 30-32). *"Little and often"* is usually a good motto for applying fertilisers. The doses for tree nurseries are generally less than those recommended to increase the yields of agricultural crops. A rough guide to start with might be:

(a) *Potting mixes:* 1-5 g of solid fertiliser per litre of soil, or 1.5 g/l for slow-release fertiliser;

(b) *Nursery beds:* 25-50 g of solid fertiliser per m^2, worked into the top 20 cm;

(c) *Dissolved in water:* 30-50 milligrams of fertiliser in 10-25 ml water per container (or 3-5 g/l in a watering can), washed off the leaves immediately afterwards.

How do I find out what suits my conditions?

Further information on fertilisers can be found in Carter (1987), Goor and Barney (1968), GTZ (1976) and Pancel (1993) in C 61-A, and from local manuals. If you think that fertilisers might be beneficial in your nursery you could do an experiment (C 15), in which for instance you might compare adding 0, 10, 25 and 62.5 g NPK per m².

Sample results: *average gain in height of experimental trees after 6 weeks.*

Treatment number	**0**		**1**		**2**		**3**
NPK added to bed (g/m²)	0.(control)		10		25		62.5
Gain in height (weeks 0-6)	3.4		15.6		16.7		13.9
Significance of differences		***		n.s.		*	

(See sheets C 55 and C 69 for assessing and analysing your results)

When should I *not* use fertilisers?

Avoid adding them:

(1) when the potting or nursery soil is rich enough anyway, for this would be unnecessary;
(2) if adding mycorrhizal or root nodule inoculations (C 30-32), because this is likely to discourage infection;
(3) to the rooting medium when rooting cuttings (A 35 in Manual 1), since for some time there are no roots to absorb them;
(4) to the germination medium (Manual 2), as seeds usually contain enough stored nutrients to last until they are transplanted;
(5) to the leaves when they are wet, for they could be damaged unless it is washed off at once;
(6) if young trees are under water stress, or are actually wilting, since fertilisers could increase the stress still further (C 41);
(7) just after transplanting or repotting, as the roots should grow into the new soil first; ***or***
(8) just before planting out, because this would stimulate rapid shoot growth at an inappropriate time (C 34, C 47).

Are there some cheaper substitutes for commercial fertilisers?

Yes, there are several locally available sources, such as:

for nitrogen: the leaves, roots, twigs and small branches of nitrogen-fixing crop plants, bushes and soil-improving trees (C 32);
for phosphorus: bone meal, made by grinding up animal bones; ***and***
for potassium: ash from recent wood or charcoal fires.

How would I use them?

For transplant beds, you could apply:

nitrogen as a compost dug in or mulch laid on the soil; by growing a green manure and digging it in; or by letting leaves of suitable shade trees or hedges fall on it;
phosphorus worked into the topsoil every year or two; ***and***
potassium as a top dressing.

For potting mixtures, an appropriate NPK supplement could be added where needed.

Micro-organisms, nutrients and tree growth C 34
- *size of planting stock*

How big should I grow my nursery trees?

This depends a great deal on:

(A) the tree species, and sometimes the particular genetic origin used (C 4-5);
(B) the conditions you grow them under in the nursery; ***and***
(C) what will be the climate, terrain and soil type of the planting site (D 11-12 in Manual 4), and what other plants will be growing there (D 14).

Do I want the maximum possible growth?

Generally no, because very vigorous growth in the nursery usually means:

(a) **larger leaves** that lose a lot of water and are easily torn (C 12-13);
(b) **taller stems** that may be inconvenient to handle and more liable to damage; ***and***
(c) **large containers** to accomodate very big root systems, which will be heavy to carry and perhaps difficult to plant (C 6, C 47); ***though***

Occasionally yes, for example when:

(1) trying to get young trees quickly to the desirable size for planting;
(2) speeding up the early stages in multiplying a new clone (A 20 in Manual 1); ***and***
(3) in certain types of experiment (C 15), when studying the potential for rapid growth.

For enrichment planting of woodland (D 24 in Manual 4) with trees such as *Pterocarpus,* success has been achieved in India by growing nursery trees rapidly in very rich soil in large containers, and planting them when 3-4 m tall.

Are there any general guide-lines?

(1) In most cases, bigger doesn't necessarily mean better planting stock.
(2) Root systems need to be more extensive than shoot systems (C 4).
(3) The greater the likely climatic stress (C 41) after planting, the better the *quality* of the young trees needs to be.

Why should there be more roots than shoots?

Because when trees go to the planting site:

(a) their stems and leaves have a waxy protection against drying out on their surfaces, and the small holes in them can be closed (C 12). Thus their structure allows them to tolerate a certain amount of water stress (C 41);
(b) the surfaces of young roots lack protection, and when disturbed they are less able to withstand drying out (C 11); ***and***
(c) unless the roots can deliver enough water to prevent severe water stress in the shoot (C 13), the newly planted tree will suffer a severe check to growth, shrivelling of leaves, die-back of shoot tips or death.

How can I control tree size in the nursery?

Primarily by choosing an appropriate sowing date for seeds (Manual 2) and the date of taking cuttings for rooting (Manual 1), so that the young trees should have grown to the desired size by the expected time of planting;
Secondarily by:

(A) choosing the size of the containers used (C 6, C 15);
(B) using a richer or a less rich potting mix, and nursery soil for beds; ***and***
(C) discouraging roots from growing out of containers, and root pruning young trees growing in beds (C 4).

Can't I alter the size just before planting?

Well, this could be done, for instance by:

(1) cutting off some or all of the foliage;
(2) pruning back the shoot system; ***and/or***
(3) breaking or chopping off some of the root system.

Isn't that a bit drastic?

Yes it is, particularly reducing the root system just then, for it is the key to successful establishment (C 4, C 11). Aim instead to control the way roots grow in the nursery rather than having to do a 'hacking' job at planting time.

Nevertheless, some commonly used types of planting stock do involve considerable pruning at the last minute (C 47).

What constitutes a good nursery root system?

One that is of a suitable size and type (C 4) to sustain the young tree:

(A) *while in the container:*
  (1) being grown on in the nursery;
  (2) transported to the planting site;
  (3) waiting to be planted; ***and***

(B) *after planting:*
  (1) without the attention previously given in the nursery;
  (2) while making new root growth rapidly, to become established; ***and***
  (3) soon being able to support new shoot growth.

Is it mainly a question of size?

Well, the **number** and **length** of the roots are two of the important points. If the root system is:
too small, it is unlikely to be able to provide enough water (C 13) or nutrients (C 14) for the young tree to survive, let alone grow well; ***but if it is***
too large, the tree may be difficult to plant properly (Manual 5).

In general, the root system should be as big as can easily be planted well.

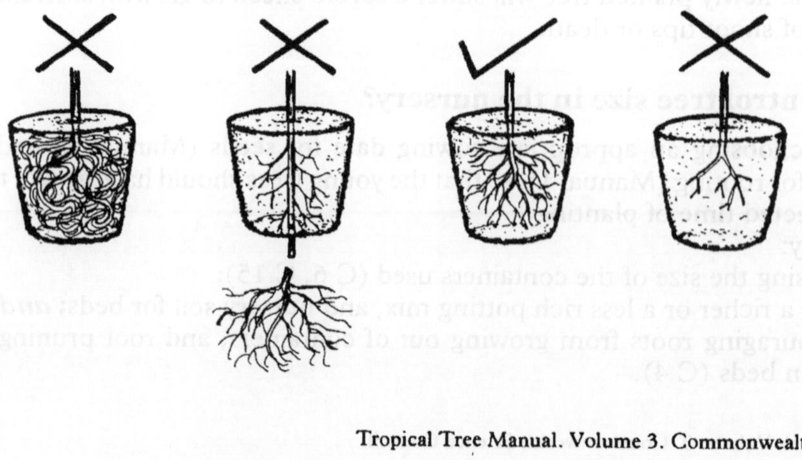

What else is important about the roots?

The **type of roots present** is as important as the size of the root system.
In the great majority of species, good planting stock needs:

(a) a lot of short roots, instead of a few long ones;
(b) a bushy root system, rather than a single taproot (C 4);
(c) many fine, absorbing rootlets as well as some thicker ones that can grow into structural roots (C 11);
(d) a compact root ball holding a good deal of soil, or a soil block, rather than bare roots;
(e) fairly straight roots, instead of ones that go round and round the inside of a container (C 6).

Inoculation with closely associated micro-organisms (C 30-32) may also make a considerable difference to how well a root system gets established.

But how can I know what is going on?

(A) By checking regularly to see if trees are rooting through from containers into the soil;
(B) By gently tapping out a few plants, if they are growing in *tapered* containers (C 6), examining the roots on the outside of the root ball, and then putting the tree back into its pot;
(C) By washing out the root systems of a few sample plants from time to time, if there are some spare trees, and the batch is not specially valuable; ***and if not***
(D) By looking at some root systems when the trees are being potted or transplanted (C 42).

Do I need to prune the roots back at this stage?

Sometimes no, for example if:

(a) potting up seedlings from shallow trays or small pots into larger containers;
(b) transferring rooted cuttings from the rooting medium to pots (A 35, A 53 in Manual 1);
(c) growing tree species that naturally form a bushy root system;
(d) handling species that respond badly to root damage; ***or***
(e) there is a serious risk of root rotting diseases (C 45); ***but***

Sometimes yes, for instance when:

(1) potting up seedlings with long roots out of a seed bed;
(2) handling older *wildings* (C 2; and Manual 2);
(3) growing species that tend to make a strong taproot and few branch roots; ***or***
(4) long roots are growing out of containers into the soil beneath. Here you could:
    (a) twist the pots round, or move them, every 1-3 weeks;
    (b) set the containers on flat stones, paving slabs or concrete; ***or***
    (c) raise them clear of the ground on wire mesh or boards.

What else could I do if the roots aren't as I want them?

You might try different:

(1) dates of potting up (C 42);
(2) types or sizes of container, for example root-trainers (C 6, C 61-B);
(3) potting mixes;
(4) amounts of shade, and rates of reducing it (C 41, C 47).

Won't all that mean a lot of work?

Not necessarily. As well as following the suggestions in manuals (C 61-A), you could:

(a) check with other tree growers as to how they obtain good root systems;
(b) compare these techniques with your existing methods in an informal trial; ***or***
(c) do a formal experiment looking at various possible modifications (C 7, C 15).

It is worth a good deal of effort to find ways of growing really good planting stock.

How big should the shoot system be?

(1) No bigger than the roots can support during the first few weeks after planting;
(2) Not too tall to be transported conveniently; ***and***
(3) Not too thin and spindly, or the plants may fall over and be blown about in the wind.

A sturdy, shortish shoot system may often be the best, even when there are weed problems in the planting site (Manual 5).

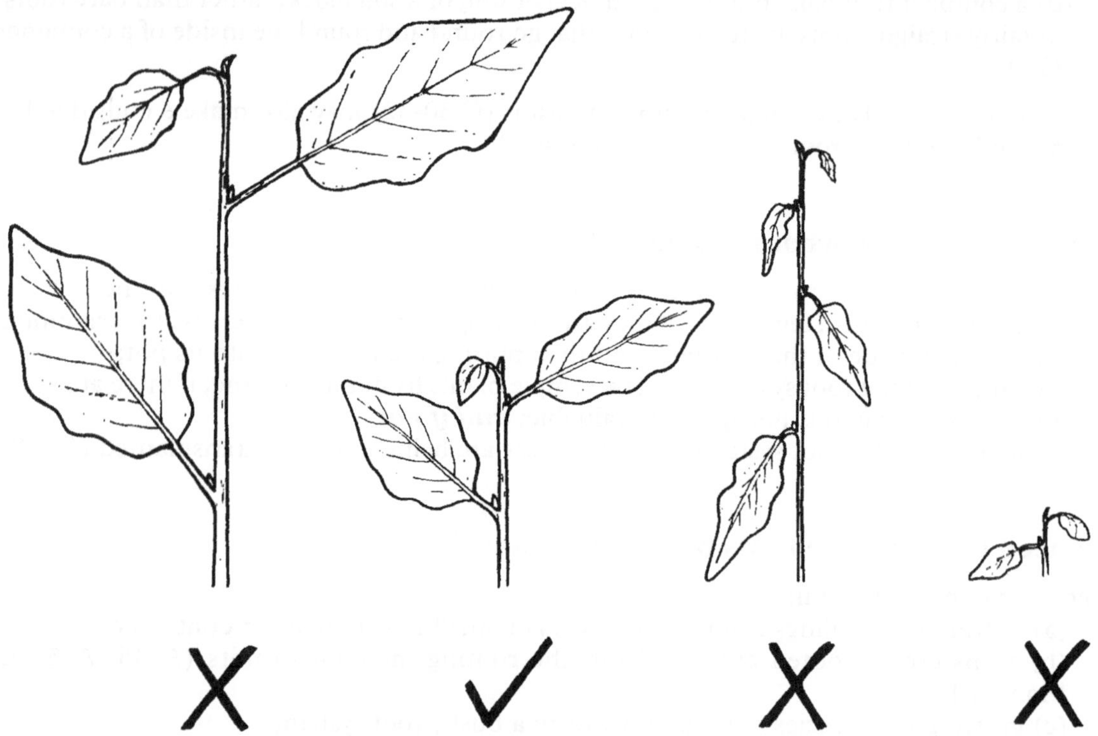

What about the leaves?

Aim for a potting mix and growing conditions that give **plenty of moderate-sized, dark green leaves.** A lot of nutrients in the potting soil (C 6, C 33) generally mean vigorous stem growth and large leaves (C 12), which may:

(a) lose a lot of water, putting the young tree under extra stress (C 13, C 41);
(b) be more liable to get damaged.

So big leaves are best avoided, except when:

(1) the number and/or size of leaves is to be sharply reduced before planting (C 47), or the trees are going to be planted when naturally leafless;
(2) growing stockplants, where the leaves will be trimmed before taking the cuttings (A 41 in Manual 1); ***or***
(3) experience has shown that they can be valuable on a particular species.

And what about the buds and branches?

In some tree species, it is important to have a good bud at the tip of the leading shoot, because lateral buds and branches:

(a) do not readily replace leaders (as for example palms and pines); ***or***
(b) grow at an angle, rather than forming a good replacement leader; ***but***

In many other cases, specific types of bud are unimportant because:

(1) new vertical leaders are produced freely from branches or the lower part of the stem after damage or cutting back; ***or***
(2) a bushy shoot habit is desired.

In species which grow by flushes, it is preferable for the buds to be inactive or growing slowly when the trees are planted out.

Does the main stem need to be straight?

Yes, this is generally best for:

(a) trees being grown for timber or raw materials (D 36-37 in Manual 4);

(b) tall shade or ornamental avenue trees (D 41); ***but***

It may not matter for:

(1) fruit trees (D 33) and ornamental shrubs (D 41);

(2) trees for forage or firewood (D 34-35);

(3) plants for hedges (C 46) or shelterbelts (D 41).

How can I keep the leader straight?

In several ways, including:

(a) selecting genetic origins that are naturally straighter (C 5);

(b) choosing a nursery site that is sheltered from the wind (C 20, C 25), and if necessary tying each plant to a small cane;

(c) avoiding over-rich potting soils (C 6);

(d) cutting out one leader if the tree forks; ***and***

(e) trimming back all but one replacement branch after damage.

What other things influence the balance between roots and shoots?

The internal control systems of the tree (C 14) are affected by:

(A) **Overhead shade:** young trees *kept* in very heavy shade tend to have elongated stems and a poor root system.

NOTE: ***this does not apply to temporary heavy shading for a week or two after potting up (C 41);***

(B) **Self-shading:** naturally bushy and fast-growing species can easily become spindly if they shade each other and their own lower leaves. Spacing out the pots is the answer, in a way that stops them falling over;

(C) **Pot size:** trees grown in very small containers can easily have more top than root;

(D) **Nutrients:** plants grown with a lot of nutrients often have a smaller root system relative to the shoot system than those in less rich soils; ***and***

(E) **Closely associated micro-organisms** that form mycorrhizas or nodules (C 30-32) may in some cases mean that the tree's shoot system is relatively larger than its actual root tissue.

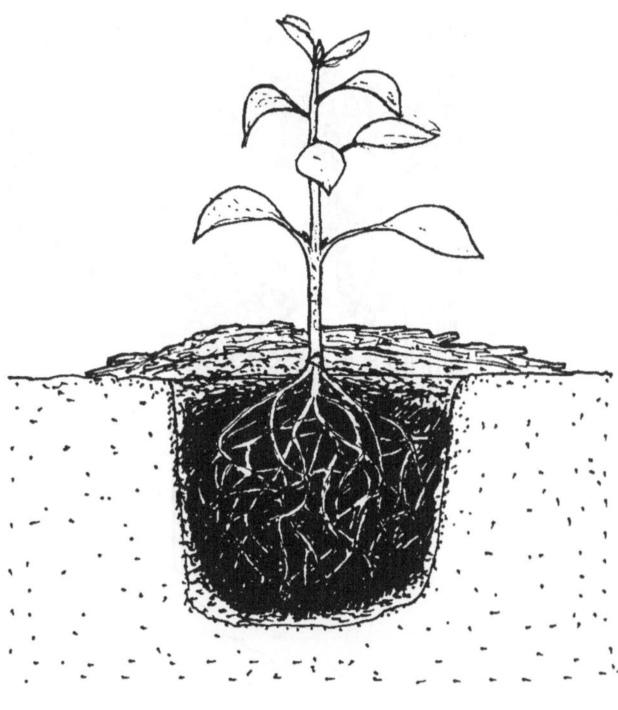

PROTECTING GROWING TREES

Protecting growing trees C 40
- *introduction: care and checks*

Do young trees really need much looking after?

Yes they do, because it's quite easy for them to be damaged or killed, especially when their root systems have been disturbed (C 2).

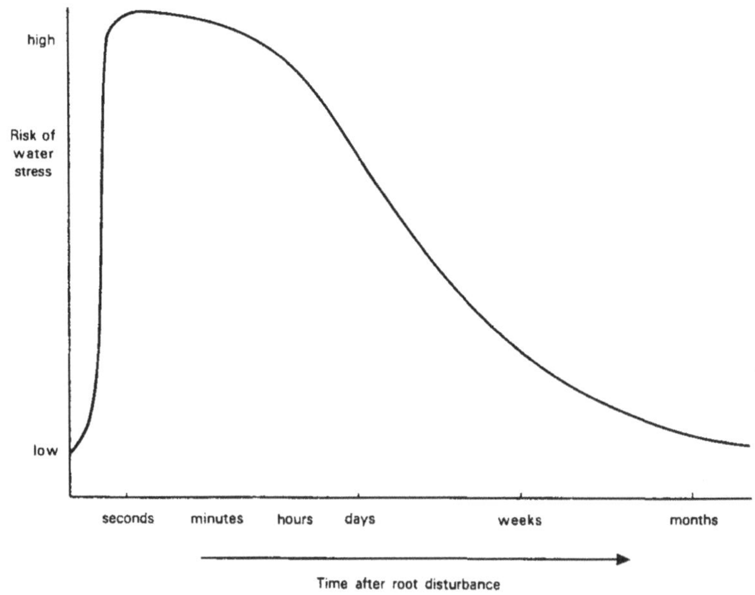

But trees in the wild have to survive without help!

That is so; but most of them don't make it.

Doesn't nursery stock need to be hardy, though?

Yes, in the sense that young trees for planting have to be able to stand up to various kinds of stress; ***but***
No, in another way, since they are ***more*** likely to survive and grow well if they are:

(A) first given favourable environments to grow in (C 10-14);
(B) protected from damage and stress (C 41), especially through key stages in their development; ***and then***
(C) *hardened* to prepare them for being taken to the planting site (C 47).

How can I provide good growing conditions for nursery trees?

This is best achieved by:

(1) understanding something of how trees grow (C 10-15);
(2) choosing the site and planning the nursery carefully (C 20-26);
(3) producing favourable environments for germination of seeds (Manual 2) and rooting of cuttings (Manual 1);
(4) preparing good soil mixtures and using appropriate containers (C 6) and nursery beds for growing on good planting stock (C 4);
(5) keeping up an adequate supply of nutrients, taking close associations with root systems into account (C 30-34); ***and***
(6) overcoming various obstacles (C 3, C 60) which can sometimes hamper nursery work.

Remember that environments that suit one species may not be appropriate for another.

Which are the key stages when special care is needed?

(a) Germination of seeds (Manual 2);
(b) Root formation in cuttings (Manual 1);
(c) Potting up (C 42) and other transplanting of young trees;
(d) Transporting planting stock (C 47); ***and***
(e) Planting, and the hours and days immediately after it (Manual 5).

Special care is needed when growing trees for research (C 7, C 48), and throughout the period of a potted plant experiment (C 15).

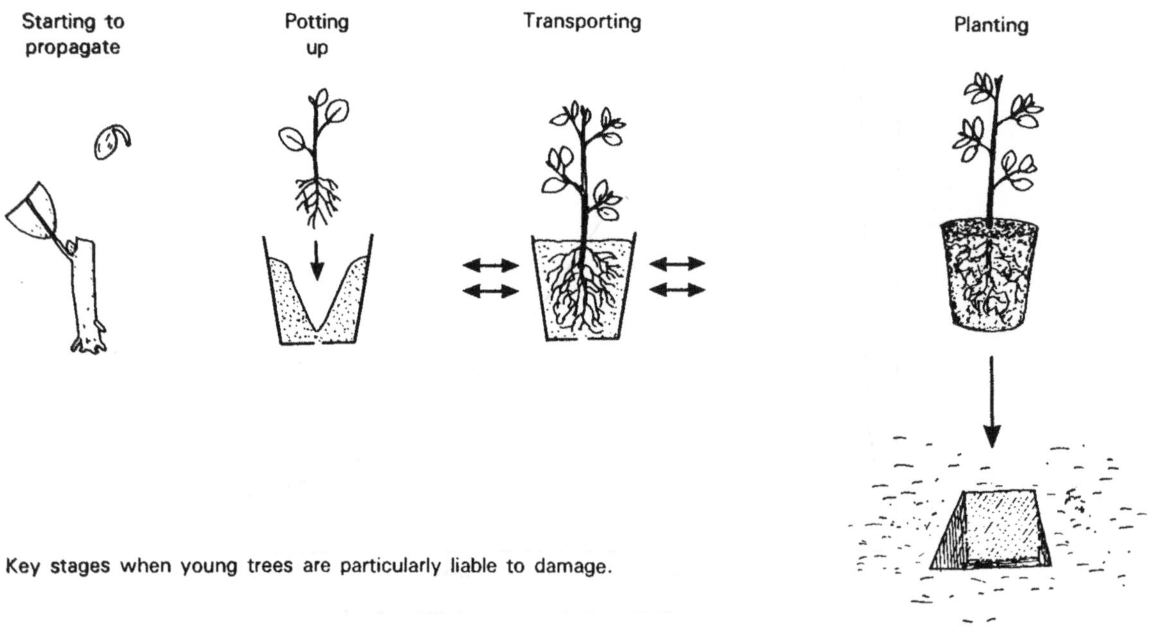

Key stages when young trees are particularly liable to damage.

Does hardening mean taking away the shade?

It usually involves **gradually** removing some shade from shade-bearing species, and all of it from light-demanding trees (C 47), but also includes:

(1) making sure that the root system is developed within and not outside containers, and that roots in nursery beds have been periodically pruned (C 6, C 11, C 34);
(2) keeping the nutrient supply adequate but not excessive (C 33-34); ***and***
(3) slightly reducing the frequency of watering or the amounts given (C 43).

What sort of checks are needed in a tree nursery?

Checks can be divided into those that need doing:

(A) before, during and after a particular job is done; ***and***
(B) regularly, throughout the year.

Which things need checking beforehand?

If you are going to do a job in the tree nursery, you need to know whether:

(a) the weather and the season of the year are suitable;
(b) there is not a more urgent job needing to be done (C 54);
(c) adequate supplies of water, materials and tools are at hand (C 24, C 51);
(d) staff and workers understand how the job should be done (C 52); ***and***
(e) enough space is available to work in, as well as for the young trees to grow in.

What kind of things need thinking about at the time a job is done?

For example, you could keep an eye on whether:

(1) the work is being done well, not carelessly; too slowly or in a rush; ***and***
(2) the young trees might be starting to suffer from climatic stress (C 41).

And which items need looking at afterwards?

You might check if:

(a) the job has been properly finished;
(b) the young trees have been left in suitable conditions; ***and***
(c) tools have been put away (C 51) and the area tidied up.

Does this apply to potentially damaging events as well?

Yes, it can do. For example, for severe storms you could:

(1) notice that one was approaching, or had been forecast, and take appropriate action;
(2) move valuable potted trees inside that were beginning to be damaged; ***and***
(3) afterwards stand up trees that had been blown over, firm in any loosened soil, check labels and remove broken shoots.

What about regular nursery checks?

These are very important indeed (C 66). They can often be **combined with the daily work,** especially if you get into the habit of looking out for:

(a) drying soil and any leaves beginning to wilt or close up during the day (C 41, C 43);
(b) water standing above the soil level in a container, or algae or moss at the soil surface, indicating poor drainage (C 6, C 11);
(c) holes in the foliage or other indications of insect pests (C 45);
(d) changes in leaf colour, premature shedding of leaves, or die-back of shoots (C 60), and any signs of infection with micro-organisms that cause disease (C 45);
(e) breaks in overhead shade, fences and hedges, or the presence of animal droppings (C 46; and D 15 in Manual 4).

How often do I need to check up?

Most tree nurseries need visiting (C 66):

(A) **once or twice every day** for watering, when experienced members of the nursery team (C 52) could also check points (a) to (e) above; ***and***
(B) **each week** for more thorough checks, and for record keeping (C 54, C 64-65).

What are some examples of weekly checks?

(1) Any tendency towards choking of beds or pots by weeds (C 44);
(2) Testing whether trees are starting to root through from containers into the ground;
(3) Indications that trees are beginning to shade each other too much, which would make the stems spindly and weak, and perhaps let the branches become entangled;
(4) Signs of major damage to roots or stems, or death of whole trees (C 60);
(5) Whether some batches of plants (C 65) need potting on (C 42) or feeding (C 33-34);
(6) Need to prune hedges, clear fire-lines or remove troublesome items;
(7) Evidence of jobs being neglected or done poorly, especially watering (C 43).

Are records really necessary?

It is usually worth keeping a simple nursery diary or notebook (C 54), because after a week or two, few people can remember just what was done, on what date, and how long it took; unless there is a written record.

With records, you will be able to:

(a) check the origin of promising batches of plants (C 64-65);
(b) avoid some problems (C 3, C 60) when growing a species for a second time;
(c) find out how long it took for young trees of different species to reach a suitable size for planting (C 34);
(d) know roughly how long a particular job should take; ***or***
(e) detect whether some plants are missing, and perhaps stolen.

But shall I really need all that information?

In one way no, as you may never look at some of it again; ***but***
In another way yes, since you generally can't tell beforehand what is going to prove valuable.

For some nurseries, sets of record sheets (C 64-67) may be convenient, while for others a brief note in a diary may be sufficient.

Are there some general hints on nursery problems?

Here are a few:

(A) Checking regularly is usually easier than correcting problems afterwards.
(B) Learning to spot the common symptoms of stress and damage is not difficult (C 60), and removing some of the causes is well worthwhile.
(C) After a time, one can learn to anticipate likely recurring problems, and take steps to avoid them.

Protecting growing trees - *avoiding various kinds of stress* C 41

What sorts of stress can affect young trees?

(A) **Climatic stress,** especially the interlinked positive and negative effects of light, temperature and moisture;
(B) **Nutrient stress,** including both too little and too much;
(C) **Mechanical damage,** due to wind, heavy rain, animals or humans;
(D) **Attack by pests** or **diseases** (C 45).

How can light harm young trees?

If they receive **too little** light for several weeks, trees can be directly damaged by:

(1) running short of sugars (C 10), so that all aspects of growth are checked;
(2) producing small, sometimes yellowish-green leaves (C 60); ***and***
(3) making elongated but weak stems.

If the light is **too bright,** even for an hour or less, trees can be indirectly damaged by high temperatures and/or moisture stress.

What harm can high temperatures do?

Indirect damage: by high air temperatures inducing very rapid loss of water from the leaves, which cannot be replaced quickly enough to prevent severe water stress and wilting (C 13). This usually interrupts growth; and can sometimes lead to browning or die-back of young leaves and shoot tips.

Direct damage: may be caused to growing cells (C 10) by air temperatures over about 40°C (D 11 in Manual 4), which can begin to disrupt their control over various chemical reactions. The tissues most liable to damage include:

(A) roots in exposed topsoil and in the atmosphere:
(B) young expanding leaves and elongating stems; ***and***
(C) dividing cells underneath thin bark.

Young trees could also be harmed or killed if they were anywhere near a fire (C 3).

Is the temperature of the soil important?

Yes. Exposed to full sunlight, the surface layers of the topsoil can exceed 40°C or even 50°C, damaging or killing the network of fine roots there, and also the mycorrhizal threads (C 31) and other important micro-organisms. Similar problems might occur if trees were watered using a supply which had become too warm in the sun (C 43).

Do low temperatures matter?

Root systems can also be harmed by the sudden drop in temperature if they are drenched with cool water in the heat of the day.

In many tropical tree species, the roots are severely damaged by temperatures of about 10°C or below, which interferes with the uptake of water (C 11, C 13) and nutrients (C 14). Such conditions could happen for example:

(a) when lowland tropical species are grown in the subtropics or in mountains;
(b) if cold water from mountain streams is used for watering; ***or***
(c) during temperature experiments in controlled environments (C 48).

Species	Minimum temperature (°C)	Maximum temperature (°C)
Acacia albida	6	42
Acrocarpus fraxinifolius	*19	*28
Albizia chinensis	-1	32
Alnus acuminata	4	27
A. nepalensis	13	26
Cajanus cajan	*18	*28
Calliandra calothyrsus	18	30
Cassia siamea	13	35
Euphorbia tirucalli	9	37
Gliricidia sepium	14	41
Grevillea robusta	6	28
Jacaranda mimosaefolia	8	34
Leucaena leucocephala	16	32
Maesopsis eminii	16	32
Morus alba	0	43
Prosopis chilensis	10	30
Sesbania sesban	10	45

From Egli and Kalinganire (1988)—see sheet C 61-D.
(* = mean annual or preferable temperatures.)

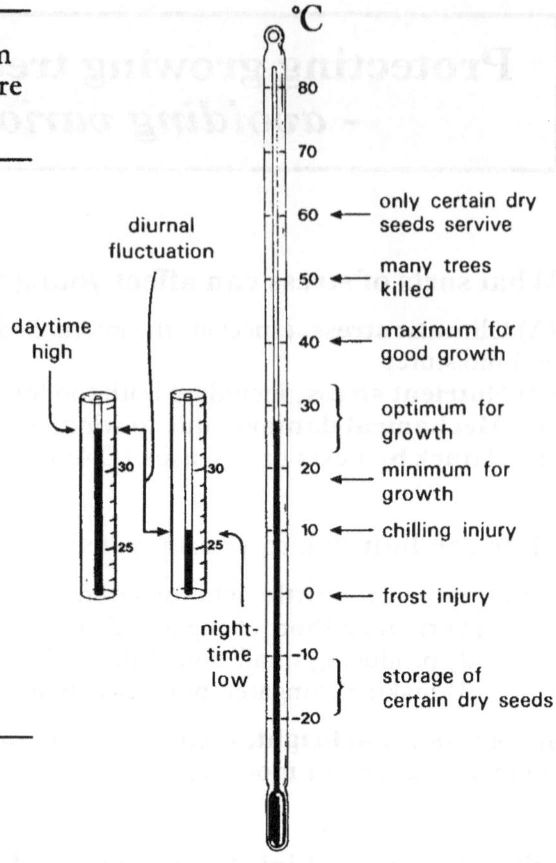

What about water stress?

Shortage of water is probably the commonest source of damage to young trees. Because we cannot see it, we often underestimate how much water is lost (C 13). For instance, a single leaf 7 cm by 7 cm could be losing as much as 1 ml an hour.

If you pull up a weed during a hot day, it may be only a minute or two before it is starting to wilt. That doesn't matter with weeds, but the same is true for most nursery trees, and a substantial check to growth can occur ***even before*** any signs of wilting.

Why is this?

Because the *guard cells* (C 12) generally close in leaves under pronounced water stress. Although this restricts further loss of water, it also means that *photosynthesis* (C 10) is interrupted (C 13). Because hormones are released (C 14), the effects of a single period of wilting may last for several days.

What other things alter the amount of water a tree loses?

In general, more water will be lost:

(1) the more **light** reaches it;

(2) the higher the **temperature** of its leaves, though above a certain point the guard cells often close;

(3) the lower the **humidity** of the air around the leaves; ***and***

(4) the greater the amount of **wind.**

Can the tree do anything else itself?

Yes, it can. For example, less water is lost when:

(a) the guard cells close during the night (and sometimes for an hour or so in the middle of sunny days);

(b) smaller rather than larger leaves are formed (C 12, C 34); ***or***

(c) the tree naturally loses some or all its leaves at a particular time of year.

Isn't too much water bad for trees, too?

Yes, stress can also be caused if all the air is forced out of the soil, leaving it waterlogged (C 6, C 23).

How can I avoid climatic stress?

The most important ways of limiting its likelihood and severity are by:

(A) **Shading** the young trees, particularly for:

(1) species that are shade-tolerant; ***and***

(2) all species during the key stages when special care is needed (C 40);

(B) **Watering** young trees regularly and carefully (C 43), weeding containers and beds (C 44), and using mulches where appropriate (C 33; and Manual 5);

(C) **Protecting** the whole nursery area against wind (C 25);

(D) **Preventing** roots from getting long instead of well-branched (C 4, C 11); ***and***

(E) **Avoiding** *sudden* exposure of young trees to more stressful conditions.

(Shading removed at the front to show the young trees.)

Which type of shading is best?

Firstly, you could leave a few suitable large trees around the nursery, and plant some smaller shade trees, shrubs or hedges (C 46).

Secondly, you will need to put up shading over the young trees themselves, which could either be:

(a) *low shade,* at a height of about 30-50 cm above the tops of the trees; ***or***

(b) *high shade,* at about 1.5-2.0 m.

How is it put up?

(A) A supporting structure is usually made out of small poles, bamboo or sawn wood, preferably extending over a somewhat larger area than the young trees beneath;

(B) The shade is placed on top, consisting of such materials as palm or banana leaves, coarse grasses, shading mats or plastic shadecloth (A 24 in Manual 1). Hanging mats may be added if needed to exclude direct early morning or late afternoon sunshine.

What is meant by nutrient stress?

There are two main kinds:

(1) a pronounced ***shortage*** of one of the 12-13 essential nutrient elements (C 14; and D13 in Manual 4), which can severely restrict tree growth; ***or alternatively***

(2) ***too much*** of one or more nutrients in the soil, so that the:

(a) levels in the soil solution are toxic to the roots;

(b) uptake of a different nutrient is hampered; ***or***

(c) water intake is restricted (C 13).

How would I know what was wrong?

Both a shortage and an excess of a nutrient can show up as distorted and abnormally coloured leaves as well as in poor height growth (C 60).

However, it is usually clear whether the soil is poor or over-rich.

What should I do about it?

For a nutrient shortage: you could add compost, mulch or a little fertiliser (C 33) to the topsoil of beds, or make up an improved potting mixture (C 6). Remove vigorous weeds, as these can quickly deplete the soil of its fertility (C 44).
For an excess of nutrients: work in some washed sand to nursery beds, and reduce the proportion of rich materials and any fertilisers used in the potting mix.
For an imbalance between different nutrients, try changing any fertiliser used, or replacing it with a suitable compost (C 33).

Do I need to know which nutrient is causing the trouble?

Generally no, because all the nutrients (C 14) are usually present in reasonable nursery soils (C 23) and potting mixes (C 6), and seldom occur to excess in them; ***but***
Occasionally yes, for instance if the local soil is:

(a) seriously lacking in a nutrient, in which case both crops and trees are likely to be suffering from the same problem;

(b) based on a type of rock that contains unusual proportions of chemicals; ***or***

(c) polluted with a chemical (such as a heavy metal like lead or mercury) that disturbs the uptake of nutrients.

If problems remain, you might need a chemical analysis of the soil to detect the cause.

Can very acid soils cause trouble?

Yes, and so can strongly alkaline conditions (C 23), except for tree species that are well adapted to the one or the other. In very acid soils, aluminium may be released in large amounts, disturbing the structure of the soil and nutrient uptake by the roots.

An acid pH may need to be changed to around neutral (C 6, C 23) for such trees as *Wrightia religiosa,* since iron in the *ferric* form makes the leaves go yellow or even whitish in colour, whereas they turn green when it is in the *ferrous* condition.

What kinds of mechanical damage can the wind cause?

Strong winds can harm young trees in several different ways, such as:

(1) clashing leaves and twigs against each other, causing shredding and breakages;

(2) loosening the soil around the root system;

(3) knocking over potted plants, and rolling them about;

(4) uprooting the tree altogether or snapping off the main stem; ***and***

(5) throwing other unfixed items around the nursery, and dropping branches on top of young trees or buildings.

How can I protect young trees from storm damage?

(A) Through careful choice of the nursery site (C 20), and by protecting it well (C 25);
(B) By securing loose items, using temporary covers or moving particularly valuable or vulnerable plants inside.

Protection from wind and birds.

Supposing there is heavy rain as well?

This can also damage young nursery trees in a number of ways, for instance by:
- (a) bending down or breaking the shoots of recently germinated seedlings;
- (b) making the foliage temporarily heavier, so that wind damage is greater, particularly in slender or brittle plants;
- (c) washing away soil so as to loosen and expose roots;
- (d) removing soluble nutrients from containers and the topsoil of seed beds; ***and***
- (e) if prolonged, making waterlogging and *damping-off* disease more likely (C 45).

What kind of protection is possible?

(A) **Permanent translucent covers,** made for example of palm leaf or shadecloth, can give some protection, especially when they are not far above the young trees. Robust covers of strong polythene sheeting can also protect, but you will then need to water the plants underneath.
(B) **Temporary opaque covers,** such as matting, boards or aluminium roofing sheets, could be used for the night or for short periods in the daytime.
(C) **Moving plants temporarily into a building,** especially if it has windows, might be worthwhile for seed trays, and for selected valuable or vulnerable trees in containers. Special shadehouses or greenhouses are sometimes used for germinating small seeds (Manual 2) and for experimental work (C 48).

In addition, shelter belts and hedges may check the force of driving rain, and free drainage of rainwater can be encouraged by raising beds above the level of paths (C 23), making sure containers have sufficient holes (C 6), and including some coarse sand and fine gravel in the soils used.

How about damage by animals?

For protection against insects, birds and other small animals, see C 45; and A 52 in Manual 1;
For larger domesticated and wild animals, see C 46; and D 15 in Manual 4;
For accidents and damage by humans, see C 3; and D 16, D 66 in Manual 4.

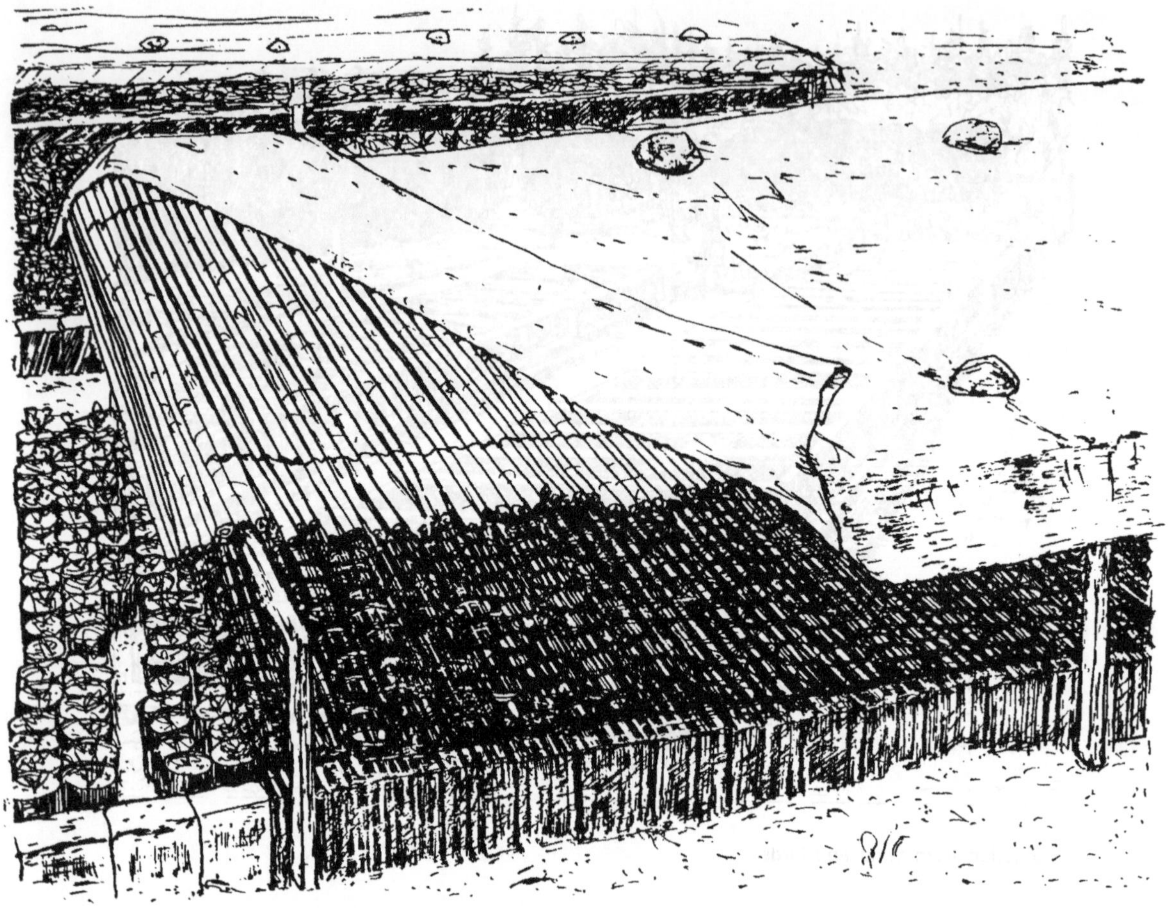

Bamboo mat shade, and polythene sheet giving temporary protection from heavy rain.

Are there any other kinds of damage?

If weeds (C 43) multiply in containers or nursery soil, there might be problems with:

(1) shortage of water and nutrients and light for the young trees;

(2) direct constriction of tree growth by twining climbers, or damage from chemical substances produced by the weed (D 14 in Manual 4); ***or***

(3) attraction of grazing animals to eat the weeds, with incidental damage to the young trees.

Protecting growing trees C 42
- *potting up and transplanting into beds*

Aren't potting up and transplanting trees easy?

Yes, in the sense that there is nothing complicated about them; ***but***
No, in that this is a critical stage when the young trees can very easily be damaged (C 3, C 40; and A 53 in Manual 1).

What is particularly important when doing this work?

(A) Choosing a time of day, and if possible a stage of shoot growth (C 12), when climatic stress is likely to be low (C 41);
(B) Minimising exposure of the root systems to sun and wind;
(C) Handling the plants carefully, avoiding unnecessary damage to the roots; ***and***
(D) Having a good potting mixture (C 6) or beds with a well-prepared topsoil (C 22-23).

Can I avoid disturbing the trees at all?

Some species with large seeds are best sown directly into containers or beds (Manual 2). Direct sowing in the field is occasionally possible (C 2), and direct planting of leafless cuttings is regularly used for certain species (A 4 in Manual 1).

With other trees, how should I decide which method to use?

(a) Find out what techniques have been used locally, and check whether or not they produce really good planting stock;
(b) Consider whether the nursery soil (C 23) would be suitable for making transplant beds (C 22), or could be improved sufficiently;
(c) See whether enough of the components to make up a good potting mixture can be obtained locally (C 6, C 63-D).

One approach might be to grow some of a batch in containers, plant others in beds, and compare the results (C 55).

When potting, should I fill up the containers with the soil mix first?

Sometimes yes *with small seedlings and leafy cuttings,* where a suitable funnel is useful (C 51) and a pointed stick can be used to make a big enough hole for the root system; ***but***
Usually no *with larger plants,* where it is better to hold the plant in the partly filled pot, and then add the rest of the soil around the roots.

How should the roots be arranged?

As far as possible, they should be spread out so that they can easily grow into a good root system (C 4, C 11, C 34). They should *not* be:

(1) needlessly bent and broken;
(2) put all on one side;
(3) twisted together like a piece of string;
(4) left going round and round.

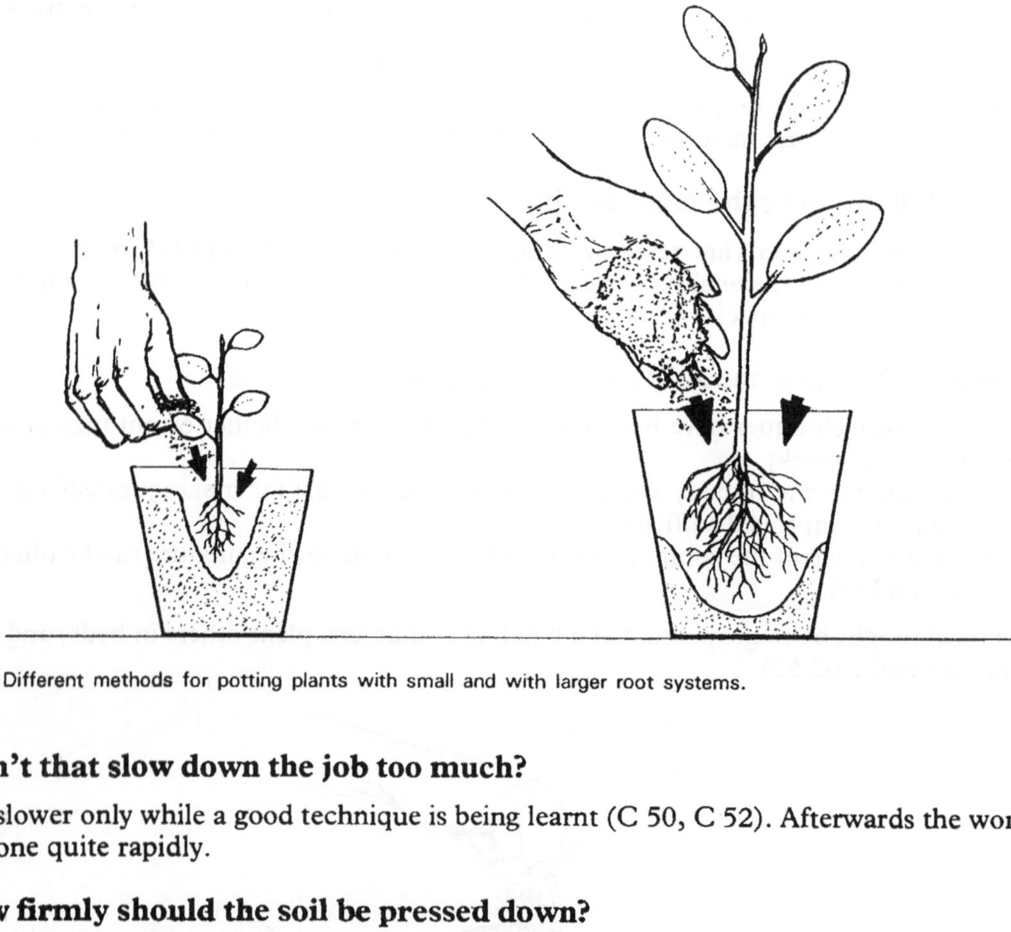

Different methods for potting plants with small and with larger root systems.

Won't that slow down the job too much?

It is slower only while a good technique is being learnt (C 50, C 52). Afterwards the work can be done quite rapidly.

How firmly should the soil be pressed down?

Moderately firmly, so that:

(a) the young tree is supported in an upright position, near the middle of the pot; ***and***
(b) water will reach all the soil in the container, rather than quickly running away through large air spaces (C 43); ***but***

Not too firmly, which might cause:

(1) breakage of a lot of roots; ***and***
(2) over-compaction of the potting soil.

What level should the soil reach?

It is usually best to end up with the soil level:

(a) 1-3 cm below the top edge of the pot, to allow for watering (C 43);
(b) at the same position on the young tree as it was previously. In seedlings, this should be at the *root collar,* the junction between shoot and root. In rooted cuttings, pot them up so that the highest root is covered by 1-2 cm of soil.

Should I water the plants in?

Sometimes no, leave it until the next day for very small or leafless plants potted into moist soil in cloudy weather; ***but***
Often yes, water the plants well when a batch of them has been potted, for example with larger leafy plants during a spell of sunny weather.

Where should I put the newly potted plants?

(A) Straight under moderately heavy **shade** (C 41) for the first 2-4 weeks after potting up, while the roots grow and branch in the potting mixture;
(B) Where they are unlikely to **root through** into the ground (C 6); ***and***
(C) With some **shelter** from the wind (C 25, C 46), so that water loss is reduced and the young trees will not blow over; and protected from heavy rain (C 41) if the trees are still very small. (See sheet C 63-B for estimating the space needed.)

Should I set the pots close to each other?

Potted plants are often put in rows close together, so that they support each other, and this also tends to stop the containers heating up and affecting the roots inside. However, lines of pots need to be separated with strips of wood or bamboo for species in which the shoots need more room if the stems are not to become spindly (C 40, C 60).

Is it advisable to stake the trees because of the wind?

For pots or beds, only do this if it is really necessary, and remember to avoid:

(1) breaking a large root when putting in a supporting cane; ***and***

(2) fastening the tree too tightly to the support, or it may strangle itself as it grows in thickness.

What differences are there when transplanting into beds?

In general, the techniques are similar. Some differences include:

(a) Potting up is best done at a table or workbench in a covered working area (C 22), whereas transplanting has to be done into the bed;

(b) Transplanting usually needs to be done in lines, while pots can be put into rows afterwards;

(c) Root pruning (C 4, C 34) can be done by lifting the pots, but requires skilled use of a pruning tool for young trees in beds;

(d) Trees in containers can be moved to a more sheltered position, whereas those in beds depend on hedges and the choice of a favourable nursery site (C 20, C 25);

Is anything else important?

(A) Make a note of the date and the approximate numbers of trees of each batch potted up or transplanted (C 40, C 54, C 65), as this record will allow you to look back accurately when:

(1) seeing how many survived and how well they grew; ***and***

(2) improving methods next time.

(B) Label the batches if:

(1) they could easily get mixed up; ***or***

(2) the trees may be used in research.

What should I do if the young trees don't grow well?

(a) Use sheet C 60 as a check-list for possible reasons, and what to do about them;

(b) Perhaps try some simple experiments (C 7, C 15).

At a later stage, one of the commonest problems is that the trees have become 'pot-bound', and suffering from nutrient stress (C 41). Generally, the roots will be going round and round inside the container, which also makes for problems when they are planted out (Manual 5).

Will I need to repot trees into larger pots?

Generally no, because it is better to use a large enough pot the first time, to avoid the extra handling; ***but***

Sometimes yes, for example when:

(1) they are pot-bound, and the root system needs to be improved (C 4, C 34);

(2) the trees are to stay in the nursery longer than expected; ***or***

(3) the particular species grows very slowly.

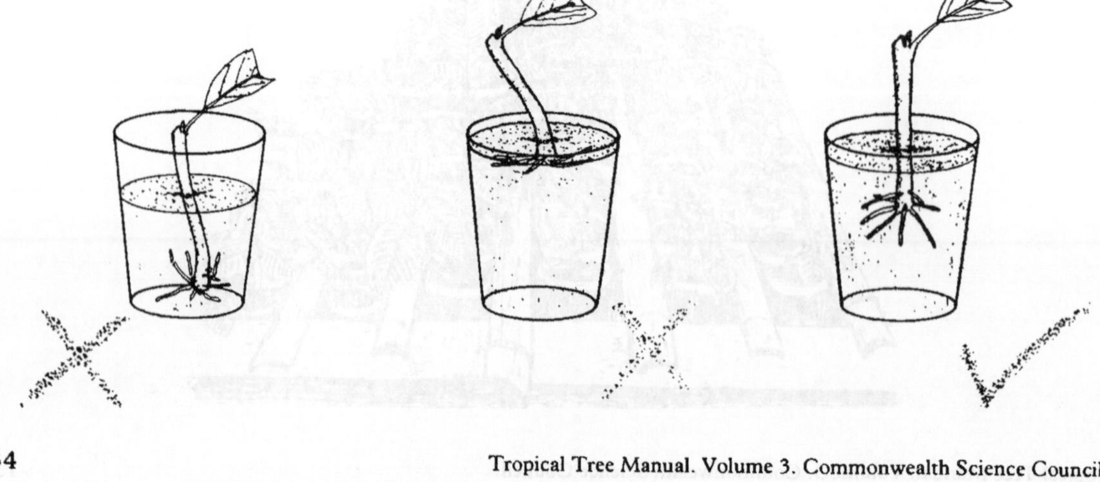

Protecting growing trees - *watering nursery trees* C 43

Isn't watering very easy to do?

Well, it looks easy, and doesn't involve difficult concepts; ***but***
There are plenty of potential problems, such as:

(A) requiring a secure water supply throughout the year, with acceptable water quality (C 24);
(B) making sure that the watering is done at the specified time, all the year round (C 40, C 66);
(C) ensuring that it is done properly, so that enough water actually reaches each root system, not too little or too much (C 41).

What can I do to avoid breaks in the water supply?

(1) Keep several big containers (for example cleaned oil-drums) filled with water as a reserve;
(2) Consider the possibility of constructing a small pond or dam to collect rainwater, catch excess from watering, or retain stream or river water;
(3) Install larger plastic or metal tanks, preferably above the level of the nursery growing areas, and put in fixed piping, taps and hose-pipes that can reach all the young trees;
(4) If the supply is pumped, then regularly clean the filters, service the motor and pump, and check for leaks and repair them (C 50).

As a last resort, you could move the tree nursery to a more suitable site (C 20).

How about avoiding breakdowns in the regular watering procedures?

The best way is to combine:

(a) an understanding by those involved of the requirements for the growth of trees (C 10-15);
(b) thorough training (C 50, C 52), with clearly explained watering procedures; ***and***
(c) a 'feel' for the watering needs of plants, rather than simply 'following the book'.

It is very helpful to have a calendar or sheet (C 66) to be ticked off when each watering has been completed. This could be double-checked by another responsible person, with special attention given to weekends and holidays.

Is it really worth bothering with all that?

Occasionally no, if the young trees are close to your home, and everyone knows what needs to be done; ***but***
Generally yes, because a day or two missed in the dry season could undo much careful work throughout the rest of the year.

What can go wrong with the watering itself?

Here are some examples:

Problem (1) - involving the types of seed trays and pots (C 6), or the kinds of nursery beds used (C 22);
Problem (2) - an unsuitable potting mixture (C 6) or unimproved nursery soil (C 23);
Problem (3) - how the potting up or transplanting was done (C 42); ***and***
Problem (4) - the watering techniques themselves.

How can they be overcome?

Here are some hints for avoiding:

Problem (1) - try out alternative containers or types of bed, on a small scale at first.
Problem (2) - change to better potting mixtures, or improve the topsoil in beds, so that they contain organic matter, but are freely drained. Then water will tend to be retained in the soil, but any excess can run through (C 23).
Problem (3) - when potting up, firm the soil sufficiently, and leave enough room at the top of the container for it to be filled up with water (about 1 cm for pots of 5 cm diameter, increasing to about 3 cm for containers 20 cm across).
Problem (4) - train staff and workers (C 52), and use more convenient methods for the actual watering, as this can often improve the standard of watering **and** shorten the time it takes. There may also be less waste of water and reduced mechnical damage to the young trees.

Well, how should the water be put on the plants?

There are various possibilities, including:

(A) buckets, with a half coconut shell, gourd or old cup to distribute the water;
(B) watering cans, preferably with a long spout and 'roses' giving a spray of water;
(C) fixed standpipes and flexible hoses, with watering 'lances' or with the ends attached to bamboo canes to make it easy to reach every tree;
(D) various irrigation techniques; ***or***
(E) automatic watering.

Do small seedlings need a different method?

Yes they do, because heavy drops of water can easily damage them (C 41, C 45, and Manual 2). Small seedlings are sometimes watered by:

(a) using a bunch of leafy twigs dipped in a bucket of water and then shaken over them;
(b) fitting a very fine 'rose', directed upwards, and passing it rapidly over them; ***or***
(c) standing seed trays in water, removing them after a few minutes, and letting them drain well.

Also, be careful about the watering of newly potted or transplanted trees (C 42).

What might interfere with water reaching the roots?

When watering larger plants, watch out for water not wetting the soil properly because:

(1) the leaves of the trees act like a 'roof';
(2) a larger tree screens a smaller one;
(3) it is difficult to reach pots at the back of a bed;
(4) the edges of polythene bag pots fold down across the soil surface;
(5) the potting soil is too loose, or has dried and formed large cracks; ***or***
(6) the work has been done carelessly.

Then isn't it best just to pour on a lot of extra water?

No, this is not advisable, because over-watering can:

(a) interrupt the supply of oxygen to the roots (C 13), which checks growth and makes root-rots more likely;

(b) increase the risk of *damping-off* disease in young seedlings (C 45; and Manual 2); ***and***

(c) wash away valuable nutrients from the soil (C 33-34).

Avoid directing a jet from a high-pressure hose at the base of a young tree, as this may expose its collar and roots, and also involve an unnecessary waste of water.

Do you mean one should add a little water often?

No, because then there could well be a dry area of soil which one can't see.

Then how can one avoid the risk of some trees getting water stress?

The best solution is to:

(A) add sufficient water to wet all the soil, with any excess draining out of the bottom of the container or soaking deeper into the bed; ***and***

(B) wait until the soil surface is getting dry, but there are no signs of wilting, before watering thoroughly again.

When should watering be done?

(a) In the early morning and/or the late afternoon, not in the heat of a sunny day (except for emergencies); ***and***
(b) Checking whether it is needed once or twice every day (C 40, C 66).

But the pots may not need watering every time!

No, that's right, because:

(1) it might have just rained heavily;
(2) the weather may have remained misty and cool all day;
(3) some of the young trees could be growing under considerable overhead shade (C 41);
(4) certain species may have few or no leaves on them for part of the time, so they do not take up much water (C 13);
(5) some containers might be large, requiring less frequent watering.

Remember to check any plants that are growing under cover, such as seed trays with delicate germinating seeds (Manual 2), newly potted cuttings (A 51 in Manual 1) or plants in greenhouses (C 48).

Isn't it a bother checking every pot?

Three kinds of watering regime can be applied:

Method (A). Pre-set times: all trees are watered at pre-arranged, fixed times;
Method (B). Conditional: each pot is checked, and only those which need it are watered; ***or***
Method (C). Pre-set/Conditional: a few sample pots are checked, and a decision then made to water all or none of the batch.

Which is the best?

Method (A) in hot, dry weather, when most pots would be getting dry later in the day.
Method (B) for small numbers of plants, valuable specimens, batches of trees for research (C 7), and especially to avoid serious water stress affecting plants during experiments (C 48).
Method (C) for tree nurseries with a well-trained team (C 52).

What is the best way of checking whether watering is needed?

(a) Looking closely for a paler colour and drier appearance of the soil surface; ***and***
(b) Testing with a finger to see whether the surface soil is becoming dry and loose.

When choosing an appropriate method, or when training staff and workers, it may be worth breaking up a sample root ball or digging into a bed to see how moist the soil is.

Cutting a test plant shows that water did not reach all the soil.

Is it different when watering nursery beds?

The basic principles are similar, though it is usually simpler to do because:

(A) it is easier to see when watering is needed;
(B) beds often dry out more slowly than containers; ***and***
(C) adequate, even, watering can be achieved more readily.

What should I do if some plants do wilt?

(1) *If the soil is dry,* water them thoroughly, and check whether they are starting to recover after an hour or two.
(2) *If the soil is moist,* **do NOT water them,** but try:

(a) repairing or increasing the shading over them (C 40-41);
(b) covering affected plants temporarily with a polythene bag;
(c) moving them for a while to a cool, shady and moist place such as a poly-propagator (A 31 in Manual 1);
(d) looking for signs of damage to the stem or root-rot (C 45) that might be responsible for the wilting.

(3) *If you have just moved the pots, or root-pruned trees in beds,* then:

(a) water the trees well;
(b) treat them as under (2 a-c); ***and***
(c) do future root control (C 4, C 34) at more frequent intervals, so the stress is less drastic.

When are irrigation methods useful?

(A) In areas with a long dry season;
(B) Where plenty of water is available; ***and***
(C) If the beds can be constructed with an even, slight slope with irrigation ditches between to carry the water.

Many methods of applying water have been installed in large tree nurseries (see pp. 133-142 in Goor and Barney, 1968; sheet C 61-A).

How about automatic watering systems?

Because these are expensive and can easily break down, they are not generally appropriate for tropical nurseries, unless:

(1) an electricity supply (mains, generators or batteries) can be guaranteed;
(2) the water supply is reliable and can be well filtered (C 24);
(3) there are large numbers of young trees, in sizeable uniform batches; ***and***
(4) the system can be adequately maintained (C 50) and is checked each day (C 40).

What else needs to be considered with automatic watering?

(a) 'Pop-up' sprinkler systems may be better than 'drip' kinds, as ants may tend to build their nests in the latter.
(b) If flexible plastic piping is used, note that this can be disturbed when pots are checked or weeded, and could be chewed for instance by mice (C 45).

Are there some other hints that might help?

(A) Stand the containers upright on level ground, even if the nursery is on a slope (C 20), so that the water does not always run to one side of the pot.
(B) Leave sufficient paths (C 22) between rows of containers, so that all can be reached easily.
(C) Water experimental plants particularly carefully (C 7, C 48), to reduce unnecessary variation between them (C 15, C 55, C 68-69).
(D) Take out weeds regularly (C 44), as some have aggressive root systems that remove water rapidly from the soil.
(E) Disturb any algae or mosses on the surface, as these make it hard to tell whether the soil is dry, moist or wet.
(F) Consider rewarding those doing the watering on the basis of how many good trees are produced.
(G) Try and reach a level of skill where other checks can be carried out while watering (C 40, C 60).

Protecting growing trees - *weeding* C 44

What makes some kinds of plants turn into weeds?

Human disturbance of land allows certain species to become weeds. They often:

(a) produce large numbers of small seeds that are distributed widely and repeatedly, and are usually long-lived and dormant (Manual 2), germinating quickly when favourable conditions occur;

(b) are colonising species, with aggressive root systems that quickly 'take over' new soil, and with rapid growth of new leaves and stems; ***and***

(c) have the ability for detached or broken parts to regenerate, so that the weed re-grows quickly after cutting, grazing or burning.

Such features allow weeds to compete successfully with young trees in cleared land.

How do they compete?

(A) By removing water and nutrients from the limited supplies in the container or bed;
(B) Because they take away growing space, and shade the leaves of the young trees;
(C) By twining around their stems; ***and sometimes***
(D) Through release of toxic substances.

Which kinds of weeds are particularly troublesome?

(A) Grasses, especially those that form clumps;
(B) Fast growing herbs and woody plants;
(C) Twining vines; ***and***
(D) Plants like *Lantana camara* that produce chemicals which can inhibit the growth of trees growing nearby (D 14 in Manual 4).

What is the easiest way of dealing with them?

Pulling or digging them out when they are still small. However, make sure you know they <u>are</u> weeds.

If they are left too long, many weeds will break off when pulled, and then rapidly grow up again from the base.

Which tools can be used to get out the whole plant?

A flat, pointed piece of cane, wood or metal of a convenient size. Smooth off the bottom to reduce damage to the roots of the young trees, and the upper end to avoid extra wear and tear on the hand.

A pair of fairly thin but strong gloves may be an advantage if some of the weeds have sharp-edged leaves, contain toxic chemicals, or are thorny like the sensitive plant *(Mimosa pudica)*.

How about cultivating the whole soil surface?

For seed and transplant beds, and for root-pruned soil blocks, the whole surface could well be cultivated before the seeds or young trees are put in; ***but***
Once the trees are growing, the soil should be disturbed as little as possible.

Is it necessary to weed containers?

Generally yes, to avoid competition with the young trees, and to remove any mosses or algae forming a crust on the soil surface, which disturbs watering (C 43); ***but***
Occasionally no, if the only weeds are small and harmless.

When should I weed?

(a) Frequently enough for the job to be easy, and to avoid much competition to the young trees; ***and***
(b) In the morning rather than later in the day, so that the damaged remnants of weeds are more likely to dry up in the heat of the day.

How can weeding be speeded up?

By doing it:

(1) when the weeds are so small they can be pulled out quickly;
(2) after watering, when the soil is softened; ***and***
(3) at the same time as other jobs, such as:
  (a) potting or repotting (C 42);
  (b) moving or turning pots to break off any emerging roots (C 40);
  (c) sorting into clones, sizes or other categories (C 7); ***or***
  (d) assessing the growth of experimental plants (C 55).

In some cases it could be quicker to move trays of plants in turn to a table at a convenient height for working.

What damage can weeding do?

If weeds are large or weeding is done carelessly, removing the weed and the soil held on its roots may mean:

(A) complete uprooting of small tree seedlings (Manual 2);
(B) disturbing and breaking the roots of older planting stock (C 4, C 11), leading to water stress (C 41) and perhaps to easier entry of disease (C 45); ***and***
(C) loosening of the soil, so water runs away without wetting it properly (C 42-43).

NOTE: since weeding usually involves putting one's hands near the ground, take steps to avoid any poisonous insects or snakes which might be amongst the containers.

How about chemical weed-killers?

Occasionally these may be needed for dealing with a serious nursery weed; ***but***
In most circumstances they:

(a) are an extra unnecessary cost;
(b) might damage beneficial soil organisms or the young tree;
(c) could leave residues that continue to pollute the environment.

NOTE: because they are very toxic, weed-killers need to be handled and applied using protective clothing, and stored in a safe place (see C 45).

What about soil sterilisation?

Heating soil for several hours can be used to kill the weed seeds in potting mixtures. Since it also kills beneficial micro-organisms (C 30), it might be avoidable by changing to a different potting mix.

If it is absolutely necessary, then:

(1) heat-sterilise only the component of the potting mix (C 6), rooting medium (A 35 in Manual 1) or seed compost (Manual 2) that contains the weed seeds or other parts that can regenerate; ***and***

(2) consider re-inoculating the sterilised soil after cooling, using a weed-free source of useful fungi (C 31) or bacteria (C 32).

Using the sun's heat to sterilise soil.

How is soil sterilisation done?

(A) **Solar sterilisation.** The soil component is spread out on the ground in full sun, completely covered with a sheet of clear polythene, and left for 2-3 days; ***or***

(B) **Sterilisation with boiling water.** The soil component is put in a metal container inside a larger one, which is then filled with water. The whole thing is set on a fire to boil for several hours or a day.

The sterilised soil needs to be kept covered and allowed to cool thoroughly before use.

Are there any other ways of counteracting weeds?

Yes, several things can be done:

(1) Cutting weeds back before their seeds ripen, outside as well as inside the nursery, including on paths and along fences;

(2) Growing tall hedges (C 46), which can reduce the entry of windblown seeds;

(3) Avoiding collecting topsoil or sand (C 24) from weedy sites;

(4) Sieving out underground parts of weeds that might regenerate;

(5) Putting mulch on beds where possible, and while they are not being used (C 33);

(6) Not putting troublesome weeds into compost or mulch; ***and***

(7) Considering retaining or planting trees that cast a light shade over the nursery (C 22, C 25), since most weeds are strong light-demanders.

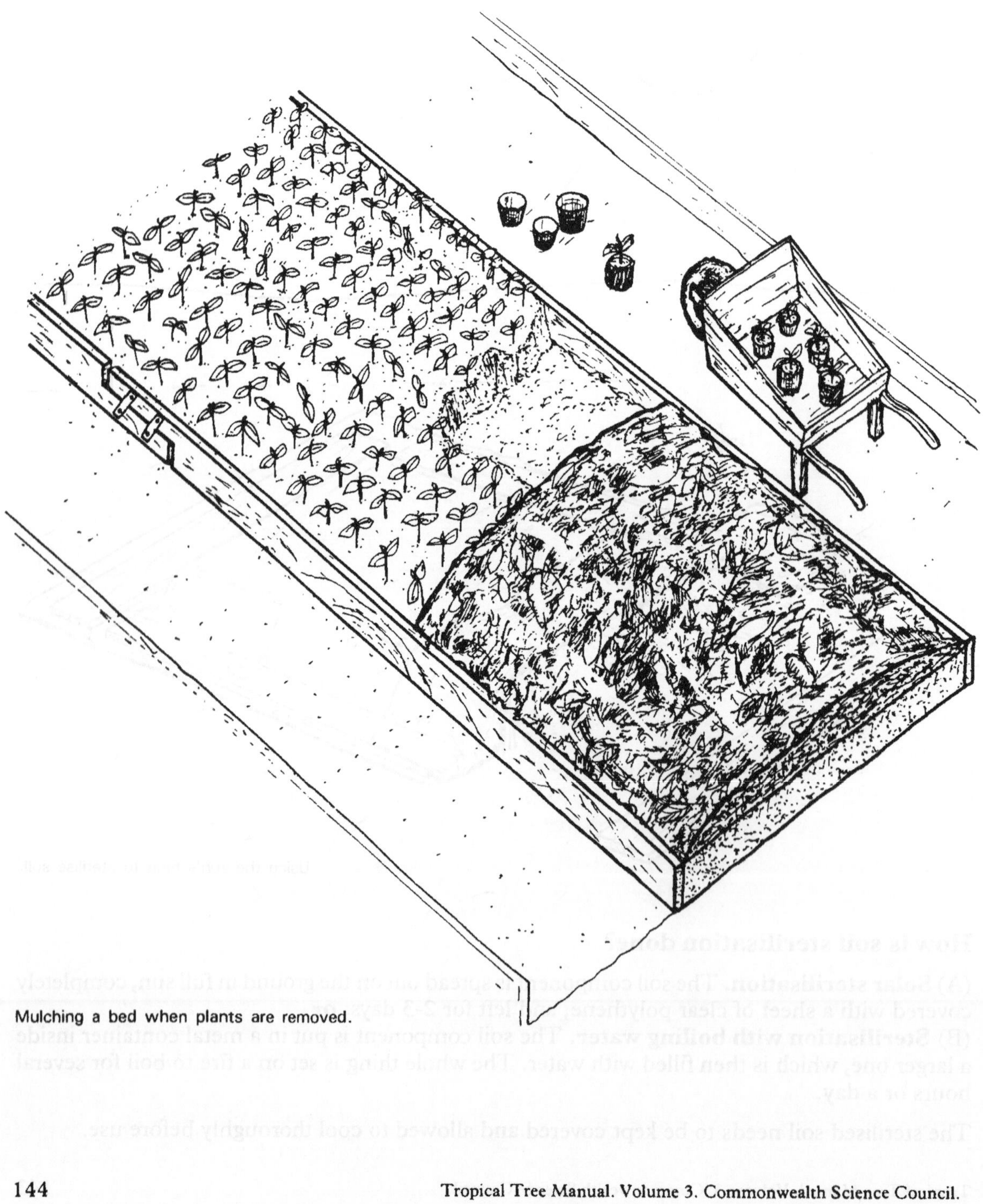

Mulching a bed when plants are removed.

Protecting growing trees C 45
- *dealing with diseases and pests*

Which kinds of disease can cause problems in a tree nursery?

Amongst those that may occur from time to time are:

(A) **Damping-off** disease of germinating seeds and young seedlings (Manual 2);
(B) **Rotting** of the root, root collar or stem; ***and***
(C) **Leaf-spots** on the foliage.

What causes them?

Certain species of *fungi, bacteria* and *viruses.* For instance:

(A) Damping-off is often caused by the fungus *Phytophthora* or by bacteria;
(B) Root-rots can be caused by the fungus *Rhizoctonia;*
(C) Leaf-spots can be caused by fungi called blights or rusts, or by viruses.

The 'spike disease' of sandalwood is caused not by a virus but by an unusual type of self-duplicating micro-organism.

How do these things damage young trees?

By getting inside and attacking some cells (C 10), disturbing their growth or killing them. In some cases the disease remains localised, but often it may spread to other nearby cells, and sometimes to different parts of the plant.

When trees are still very young, diseases can often be fatal, and may also spread easily from one seedling to the next.
Older trees could still be damaged and their growth checked, but they are usually more likely to survive.

Is there anything I can do about diseases?

Yes, there are several things. They are much more likely to occur when the conditions for tree growth are poor, so try and avoid:

(1) poorly-drained potting mixes (C 6) and nursery soils (C 23);
(2) over-firming of the soil when potting up or transplanting into beds (C 42);
(3) too heavy a shade (C 41);
(4) frequent over-watering (C 43);
(5) not enough air circulation around the young trees;
(6) seedlings growing too close to each other.

These either lead to waterlogging of the soil, or to very high humidity of the air close to the young trees, both of which tend to encourage diseases.

What else could I do to avoid problems?

(A) Learn the symptoms of common tree diseases (C 60, C 61-C), and look out for any signs of them during regular checks (C 40, C 66);
(B) Water early in the morning, rather than at the end of the day;
(C) To reduce the risk of damping-off, try watering seed trays from below (Manual 2), and don't leave them soaking for more than 10-15 minutes;
(D) To reduce the risk of root-rot, improve the drainage beneath containers, if possible raising them clear of the ground, for instance on wire mesh;
(E) Avoid contaminating the potting mix, compost or mulch with soil or vegetation from young trees that have been attacked by a serious disease, because its spores or other stages may persist in such material.

But what should I do if disease starts to attack my trees?

(a) Isolate trees showing symptoms of disease from the rest of the batch, with identity labels as needed to avoid later confusion (C 54);
(b) Collect and burn damaged leaves that have been shed and plants that have died;
(c) Keep a close daily watch (C 40) to see whether the disease is spreading or diminishing;
(d) Use a disinfectant regularly, and consider heat-sterilising the potting soil;
(e) If really necessary, spray the affected plants with a *fungicide.*

Spraying wearing protective clothing.

Why not spray them all anyway?

Because using a fungicide:

(1) may not be necessary;
(2) would be expensive to continue doing to the whole batch every time you saw a diseased leaf; ***and***
(3) could kill valuable micro-organisms in the soil (C 30-32).

Another reason is that organisms causing disease can sometimes build up resistance to the chemical, and then you might have nothing in reserve against a serious attack.

Which items would need disinfecting?

Anything that can carry the disease to a new set of young trees, such as:

(a) tools (C 51) used in digging up or handling diseased plants;
(b) containers (C 6) that are to be re-used;
(c) tables or benches used for potting up (C 42).

If the disease is persistent, also wash boots and treat the floor of the potting area.

What's the best disinfectant?

A suitable one that may be readily available is a 2% solution of ordinary domestic bleach. Add 200 millilitres of concentrated bleach to 10 litres of water, and then rinse off with water after disinfecting. Avoid getting bleach on your hands or in your eyes, and wash yourself well with plenty of water afterwards.

How is soil sterilised?

(A) *For seed and potting composts:* in the same ways as to kill weed seeds (C 44); ***or***
(B) *For micropropagation* (A 5 in Manual 1), when transferring plantlets from the medium to soil: in a laboratory *autoclave,* where the soil is heated under pressure at a high temperature for a short time.

Doesn't heat-sterilisation kill everything living in the soil?

Yes, if the temperature stays above about 75°C for an hour or so, diseases, pests and weeds (C 44) are likely to be eliminated. However, the disadvantages are that:

(a) the valuable micro-organisms that form close associations with tree roots (C 30-32) will die too;
(b) the useful *decomposers* that break down dead organic matter and release nutrients (C 14; and D 13 in Manual 4) are also killed; ***and***
(c) it is tedious sterilising large quantities of potting soil.

How soon will such organisms recolonise the soil?

The spores of fungi and bacteria will start reaching the soil as soon as it is exposed, but for some time there will many fewer than before. The risk of disease will be reduced, but so will the chance of effective close associations with beneficial micro-organisms.

Which chemicals can stop disease spreading?

There are many different substances that are strong fungicides, each of which may be given different brand names by the company selling them. One active ingredient may be effective against some disease organisms, but not against others. Some examples are:

(a) *Propamocarb* which deals with some of the common damping-off fungi;
(b) *Iprodione* is effective against fungal leaf-spots;
(c) *Benomyl* and *sulphur* are more general fungicides.

Note that the first three of these chemicals act *systemically;* that is, they are taken up in solution by the roots, and are then transported within the young tree (C 14).

What is the best way of spraying?

(1) Obtain an appropriate fungicide and a few small hand-sprayers;
(2) ***Read the instructions on the fungicide packet carefully!***
(3) Choose a windless day, or take the potted plants inside a shelter;
(4) Put on appropriate protective clothing (plastic or rubber gloves, eye-shield or face-mask and a long-sleeved shirt);
(5) Make up a sufficient amount of the fungicide, and put it into the sprayer;
(6) Spray the plants according to the instructions;
(7) Wash out the hand-sprayer where the fungicide can do no harm, and leave it to dry;
(8) If necessary, repeat the spraying after 2 weeks, if possible with another fungicide that has a different active ingredient.

A *soil drench* is an alternative way of applying systemic fungicides.

How about nursery pests?

Like fungi and bacteria, most of the very small animals that live in the soil are harmless to trees, and many are useful decomposers. However, certain kinds of insects can cause root damage, or death of young trees, particularly if their numbers build up. Similarly, some insects attack the shoots and weaken the young tree.

Which sorts can be troublesome?

(1) **Plant-sucking insects,** like aphids (greenfly), psyllids, thrips, mealy bugs and scale insects, withdraw food from the sugar-conducting cells (C 10), and can sometimes infect the young tree with a virus disease.
(2) **Leaf-eating insects,** such as caterpillars, grasshoppers, weevils and some kinds of beetles, eat young leaves and stems. Large swarms of locusts are especially damaging.
(3) **Shoot-boring insects** harm trees when they attack the leading shoot, as in *Milicia (Chlorophora)* and many plants of the mahogany family.
(4) **Cut-worms** (the young stage of a kind of moth) may break off young seedlings just above the soil level at night.
(5) **Termites** can sometimes attack young nursery trees, and they may also bore into and destroy seed boxes, planks, wooden posts and buildings.

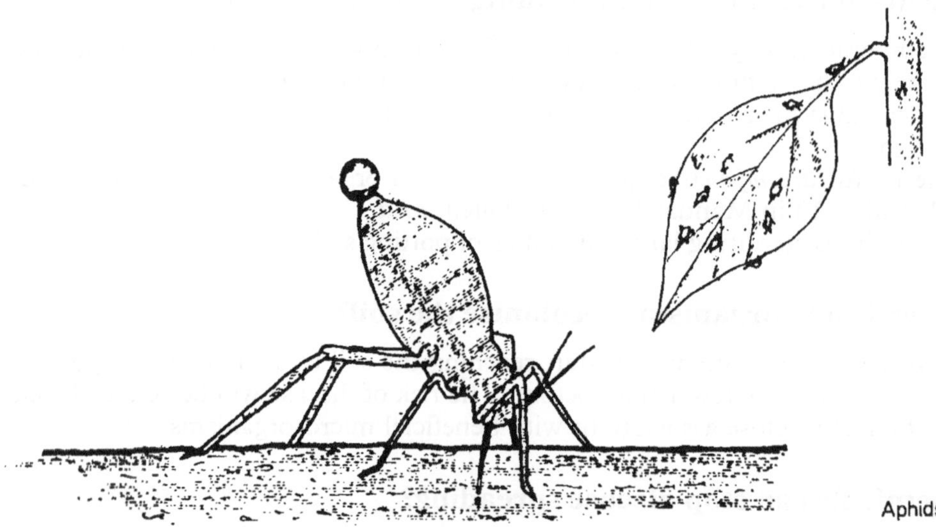

Aphids

How can harmful insects be dealt with?

(a) By looking out for any signs of them during daily and weekly checks (C 40, C 66), particularly during seasons and at stages of tree growth when attacks are more likely;
(b) Through removing or squashing any groups of insects that are starting to build up on the young trees or on nearby vegetation, or by spraying them with water containing a little detergent;
(c) Through disinfecting tools and the working area, as this will kill the eggs of many pests, and by removing dead leaves and other materials where insects might hide; ***and if necessary***
(d) By applying an appropriate *insecticide.*

Locust swarms can only be tackled on a district and regional scale.

Which insecticides can be used?

Sometimes local plants may be useful for this. For instance, aphids can be controlled by pounding up the fruits (and especially the seeds) of the neem tree *(Azadirachta indica).* Mix with water and splash the liquid over the affected plants (C 61-C).

What kinds of sprays can be bought?

As with fungicides, there are many different insecticides, which are effective against various pests. For example:

(A) *Pirimicarb* and *Dimethoate* deal with aphids;
(B) *Nicotine* is effective against thrips and some caterpillars;
(C) *Aldrin* can be used to protect young trees against termites.
(D) *Pyrethin* plus *piperonyl butoxide ('Pyrethrum')* forms a more general insecticide, which works by contact with the insect. It is also:
(1) less toxic to human beings; ***and***
(2) unlikely to remain in the nursery environment as a pollutant (D 16 in Manual 4), as it is quickly broken down in sunlight.

Are most insecticides poisonous, then?

Yes; and it is therefore ***particularly important to:***

(a) ***read the instructions carefully;***

(b) use protective clothing (rubber gloves and a long-sleeved shirt, long trousers and rubber boots. A hat, an eye-shield and a cloth tied across the mouth may be needed for insecticides that are very poisonous to humans;

(c) avoid unnecessary spraying;

(d) dispose of unused insecticide thoughtfully, and wash out the sprayer carefully;

(e) label insecticide containers clearly, avoiding any confusion with those containing drinks, and keep them locked away; ***and***

(f) wash hands before eating, drinking or smoking.

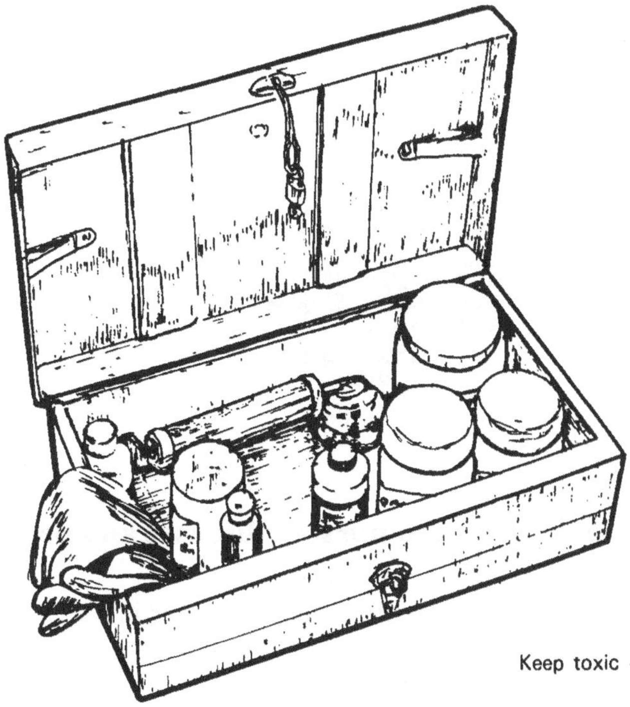

Keep toxic chemicals locked up!

Will soil sterilisation kill insect pests?

Yes, if the temperature exceeds about 75°C for an hour. On a small scale you could try pouring boiling water on to insects or their eggs, provided that any young trees are at least 0.5 m away.

What other pests can cause problems?

(A) **Nematodes,** which are small thread-like soil organisms that can attack the roots;
(B) **Small rodents,** such as mice and rats, which can eat, break off or disturb young trees;
(C) **Seed-eating birds,** which can dig up and eat newly sown or germinating seeds.

How can I deal with them?

(A) Nematodes can be treated by drenching the soil with a *nematocide,* if necessary. *Carbofuran* is effective, but is **very poisonous.**
(B) Rodents can be checked by wire netting, and by trapping or poisoning.
(C) Birds can be kept off by:

(1) spreading netting over the seed beds or seed trays until well after germination;
(2) using people or gadgets to scare them away.

Is there anything else I can do about diseases and pests?

(A) One cannot eliminate them entirely, so concentrate on keeping the risks low.
(B) Try scattering individual *Milicia* and mahogany amongst other species.
(C) Avoid growing only one tree species in a large nursery, as a disease organism or an insect pest could build up more easily, increasing the risk of losing all the plants.

Protection from wind and birds.

Are some trees more resistant to nursery diseases and pests?

In each tropical region there are likely to be some:

(a) **tree species** that are seldom attacked, and others in which the risk is higher;

(b) **provenances** and **individual trees,** within a species, with greater or lesser tolerance or resistance to a specific disease or pest.

In time, superior trees could be identified, multiplied and tested. Then it might be possible to grow that species even where there are persistent nursery problems (C 3-5).

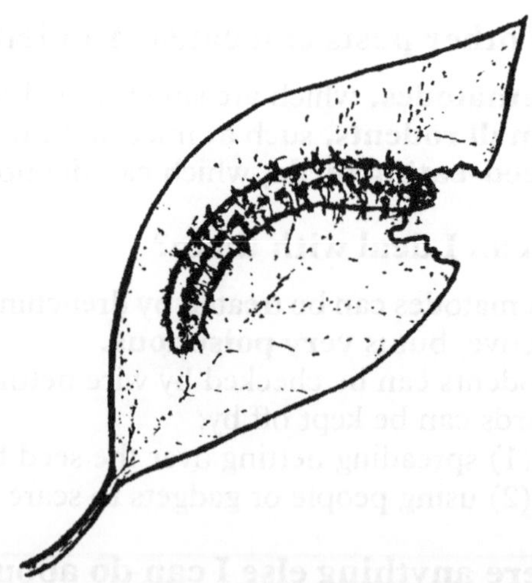

Canker.

Protecting growing trees - *fences, hedges and gates* C 46

What are the main reasons for having fences?

They are often needed to keep out:

(a) **domesticated animals** that could eat or break the young trees, for example cattle, hens and especially goats;

(b) **wild *herbivores*** (D 15 in Manual 4); ***and/or***

(c) **humans,** especially thieves and vandals (D 66 in Manual 4).

Fencing can also be used to support lines of young nursery trees in windy sites.

Does a nursery always need fencing?

Temporary nurseries (C 21) are often left unfenced, especially if there are few animals in the neighbourhood, or if they are near the house where someone can keep an eye on them;
Permanent nurseries, small or large, often need protection, although sometimes it may be unnecessary;
Research nurseries generally need to be fenced to protect the experimental trees. This is essential if the trees are being treated with radioactive or other toxic substances, where only the research personnel should have access.

What sort of fences could be used?

There are many possibilities, such as:

(1) Palm fronds, close together, with their bases inserted into the ground;
(2) Wooden posts, with wooden palings and cross-beams;
(3) Wooden posts with strands of wire between them;
(4) As in (3), with wire netting or chain-link fencing attached;
(5) Electric fences.

Support against the wind for a line of nursery trees can be provided by wooden posts with horizontal poles.

How can I get hold of chain-link fencing?

Simple fence-making machines are now available, which themselves can be built locally (C 61-D).

Supposing there is only a little money for buying fencing materials?

It is still possible to build a fence round your nursery plants, using palm fronds or small poles, split bamboo or pieces of sawn wood. These need to be put in close to each other, and interwoven or held together with lianes, rope or wire (D 38-39 in Manual 4).
However, such fences perhaps might:

(a) not last very long;
(b) blow down in strong winds; ***and***
(c) not keep out certain kinds of animal.

What can I do to make a fence last longer?

(1) Use durable wood, or treat it with a preservative against rotting and termites;
(2) Make sure that some pieces go well into the ground, and are fixed firmly to anchor the fence;
(3) Close up any holes that appear, particularly at the bottom, by weaving in small branches or pieces of wood; ***and***
(4) Plant grass to hold the soil at the base, keeping it cut to avoid increasing the risk of fire.

How tall does it need to be?

For cattle and many wild animals, 1.5-2 metres might be enough.
For goats, hens and antelopes, 2.5-3.5 metres may be needed.

How could I build a strong fence against animals?

(a) Align the fence to make use of any buildings, gate-posts and existing trees that are available, making the sides as straight as you can;
(b) Dig holes for main posts every 15-20 m, placing them where the direction changes, and using a metal crowbar or pickaxe to put them 30-60 cm deep;
(c) Fix these posts firmly by hammering stones into the hole, and by bracing them with a slanting pole on either side, or at the middle of the angle where there is a change of direction. If you concrete them in, wait 2-3 days before stage (e);
(d) Drive in smaller, pointed stakes every 1.5-2.5 m, using a heavy hammer;
(e) Starting from a main post, attach lines of fencing wire at vertical intervals of about 30 cm loosely to the smaller posts, and then strain the wires as tightly as you can to the next main post (with wire-strainers if available) before fixing them all firmly;
(f) Now attach wooden palings, wire netting or chain-link fencing securely to this strong structure, going about 3-5 cm into the ground to discourage burrowing animals.

What problems might I have in putting up fences?

(1) **Steep ground** can make it difficult to get posts vertical and well fixed, and the fencing materials might need to be cut on the slant;
(2) **Rocky patches** can mean problems in making holes for the fence posts, though once made the fence may be stronger;
(3) **Soft soil** can cause difficulties in fixing posts, unless they can be made longer and driven into firmer soil beneath.

How about animals that might cause special problems?

(A) **Animals with strong teeth,** such as mice, rats, porcupines and some insects, may eat holes in the fence. Metal fences are more resistant than wood, but may not be immune to damage.
(B) **Climbing and jumping animals,** such as monkeys, squirrels and sometimes even goats, can be very difficult to keep out, especially if there is little to eat outside the nursery. You could try having someone to scare animals, secure tethering of goats, or covering the entire area with shadecloth (C 41; and A 24, A 33 in Manual 1), with the strips securely joined together.
(C) **Large, strong animals,** such as cattle and elephants, can push over a fence unless it is very strong. Battery-operated electric fences could be a relatively cheap solution, or you might even dig a broad ditch around the nursery, as they may be unwilling to cross it.

Are there any other alternatives to fences?

Yes; planting tight, often thorny hedges, sometimes called "live fences".

What kinds of woody plants are best for that?

Those which:

(1) are in use locally, and have been found effective and robust;

(2) have a habit of shoot growth with a lot of strong low branches, which do not lose all their leaves at once (C 12);

(3) can respond to repeated pruning by producing new shoots near the ground; ***and***

(4) are easily propagated, especially from leafless cuttings (A 4 in Manual 1), like *Euphorbia tirucalli* and species of *Baphia, Ficus* and *Spondias.*

How close together should they be planted?

If there are plenty of plants or material to make leafless cuttings, you might try planting ***two*** rows, 10-15 cm apart, with the hedging plants 3-6 cm apart in the rows, and *staggered,* so that the second row fills the gaps in the first.

However, a spacing of 25 cm is recommended for some species, so first check what has succeeded locally.

What height should hedges be?

As a rough guide, start trimming the tops at 1-1.5 m, to encourage bushy growth.

Do hedges need much maintenance?

They need regular attention, especially:

(A) **Pruning** at intervals, to stop them getting too tall and/or too wide; and to encourage strong new growth from near the base of the hedge; ***and***

(B) **Replenishing** at the bottom with new plants, if parts have been broken.

Plants like Sisal can form an effective, low-maintenance barrier.

What other effects can hedges have on the nursery?

(1) Checking the run-off of water and the erosion of soil (D 23 in Manual 4);
(2) Lessening wind speed over the young trees, and so;
- (a) reducing the amount of water the young trees lose (C 13, C 41);
- (b) decreasing the risk of direct damage (C 25); ***and***
- (c) discouraging the spread of wind-blown weed seeds (C 44).

Hedges might also cast some shade on the young trees, which may or may not be an advantage.

How about access to the nursery?

It may be best to have only one gate to the nursery, at the most convenient point in the fence or hedge (C 22), because:
- (A) gates are potential weak points in the line of protection;
- (B) it is difficult to stop small animals from getting under them; ***and***
- (C) stiles can be built if people need to enter from another side.

It may also be easier to discourage vandals and thieves if there is only one padlock.

What else could I do to keep out such people?

You might try a combination of:
- (a) high fences and notices;
- (b) training staff and workers (C 52);
- (c) discussions with the local community (C 53; and D 5 in Manual 4); ***and***
- (d) employing guards at night and during weekends and holidays.

Are there some hints about putting in gates?

(1) If possible, choose a flat place, so that the gate will not have to be slanting;
(2) Decide whether a wooden or metal gate will be best, and which way will be most convenient for it to open;
(3) Use a very strong, durable main post to hang the gate on, and another post for the closing device;
(4) Fit the hinges so that the gate will hang slightly high, as gates always settle a little after a few days or weeks;
(5) In order to exclude small animals, attach some trailing strips of wood or folded wire netting to the bottom edge of the gate; ***and***
(6) Consider fitting a device to prevent the locked gate being opened simply by lifting it off its hinges.

How do I maintain a gate?

By treating it with a wood preservative or paint, oiling the hinges occasionally, and by mending it if it becomes weakened or broken (C 50). You could also fill in any ruts in the road beneath the gate with crushed stones, to continue keeping out small animals.

Is there a cheaper alternative to a gate of this kind?

Yes, you could make a separate piece of fencing that fits between double poles on both sides, but can be lifted out.

See Fearnside and Drew (1977) in C 61-D for more information on gates, fences and hedges.

Protecting growing trees C 47
- *preparing young trees for planting*

What is needed before young trees are sent out from the nursery?

Giving them the best chance of:

(A) coming through the journey with minimal damage and stress (C 41);
(B) tolerating poorer conditions while waiting to be planted (Manual 5); ***and still***
(C) surviving and growing well after planting, and becoming quickly established.

But what can be done about such things in the nursery?

Quite a lot; especially by:

(a) planning to grow the young trees to a suitable size (C 34);
(b) progressively modifying shading and watering regimes to *harden* them;
(c) continuing regular pruning of the roots (C 4); ***and sometimes***
(d) last-minute alterations to their size and leafiness.

How should shading be changed?

Young nursery trees need enough shade at critical stages (C 40). Then:

- *for light-demanding species* (D 14 in Manual 4): shading should generally be reduced **gradually in stages** every 2-4 weeks, so that they have been hardened to grow in full sunlight for at least a month before they leave the nursery.
- *for shade-bearers:* shading should be similarly reduced, but the hardening taken **only to a level of shading** suited to young trees of that species.

Does the intended planting site not make some difference?

Yes, it can do. For example, during the last month in the nursery, you might:

(1) keep even the light-demanders **under some shade,** if the site will be a shady one. Examples could include various situations where one or more *storeys* of tree crowns already exist overhead (D 52 in Manual 4), such as:

(a) *underplanting* in uniformly thinned plantations (D 51);
(b) growing trees in small groups or narrow strips in woodland (D 24, D 54);
(c) certain agroforestry situations (D 3, D 21); ***and***
(d) some ornamental and park planting (D 28).

(2) consider **removing most or all the shade** even from shade-bearers, if they have to be planted in large, open clearings (D 2, D 50) or on degraded land (D 22).

Why is this?

Because the leaves of trees (C 12) are usually *adapted* to the conditions prevailing at the time when they expanded. The new leaves which grow during the last month in the nursery will be those that provide most of the sugars which will sustain the tree during the first month after planting. So they need to be adapted as far as possible to the conditions that will apply in the planting site.

What changes should I make to watering regimes?

Continue to give plenty of water each time, but reduce the **frequency** of watering, for instance:

(a) from twice a day to once a day;
(b) from once a day to once every 2-3 days; ***or***
(c) from pre-set times to a conditional regime (C 43).

However, don't take this to the point where the young trees wilt, or come under severe water stress without showing it (C 13, C 41).

Will less frequent watering improve the chances of survival?

Yes it may, particularly by:

(A) discouraging very rapid extension of shoots (C 4, C 12), which might be more liable to damage; ***and***

(B) encouraging smaller leaves that are likely to lose less water, thus helping to preserve the water balance (C 13, C 34) of the young tree during its transition from nursery to the field.

Can I avoid watering them beforehand to make them lighter to carry?

Generally no; because:

(1) the roots will usually be less damaged by travelling in moist, firm soil than if it dries and becomes loose; ***and***

(2) you may not know when the young trees will be getting any more water; ***but***

Sometimes yes; when they are going to another nursery or to a planting site with water close by, and someone to look after them.

How about adding fertilisers?

Usually none should be added at this stage, because they are:

(A) likely to encourage too vigorous growth and large leaves (C 34) at an unsuitable moment; ***and***

(B) more likely to be washed away and wasted than if applied when the tree is potted up (C 42) or planted (Manual 5).

If nutrient stress (C 41) is severe, you could for instance:

(a) insert a little slow-release fertiliser with high P and K (but low N) into the root ball;

(b) apply a micronutrient (C 14) that was needed, perhaps by foliar feeding; ***or***

(c) delay planting out until the trees have recovered.

Why does root pruning help?

Because **regular, moderate pruning** encourages the formation of a good, bushy root system (C 4, C 11, C 34) that will go into the planting hole (Manual 5) with little disturbance.

NOTE: **this is crucial to successful growing of planting stock in containers and in soil-block beds.**

But aren't trees sometimes planted as 'stumps'?

Yes, a few tolerant kinds, such as teak, *Gmelina* and *Albizia chinensis* can be grown quite large, and then both root and shoot cut right back to a few centimetres.

However, with most species it is best to avoid getting into a position where you need to chop off big roots.

What about cutting back the shoots?

This can sometimes be a useful method, if:

(a) the trees have a good root system, but are rather tall; ***and***

(b) there are buds at the base, and adequate reserves, for the shoots to regrow.

Conifers such as *Pinus* should not be cut back, as they seldom have any basal buds. On the other hand, cypresses such as *Cupressus* or *Callitris* might be able to re-grow from green branchlets near the base.

Couldn't I just take the leaves off and plant 'striplings'?

Often, it is better to avoid this technique, because the young trees lose not only the materials and energy used to make the leaves (C 10, C 12), but also most of their ability to manufacture sugars. (In one trial, **removing all the leaves** was actually found to kill healthy plants of *Triplochiton scleroxylon,* even though they were not taken out of their containers.)

Sometimes, however, **reducing the amount of foliage** is helpful, because this decreases water stress (C 13, C 41) during the period before and after planting.

Occasionally, it is unnecessary, because the nursery trees naturally lose some or all their leaves or leaflets at the right time for planting.

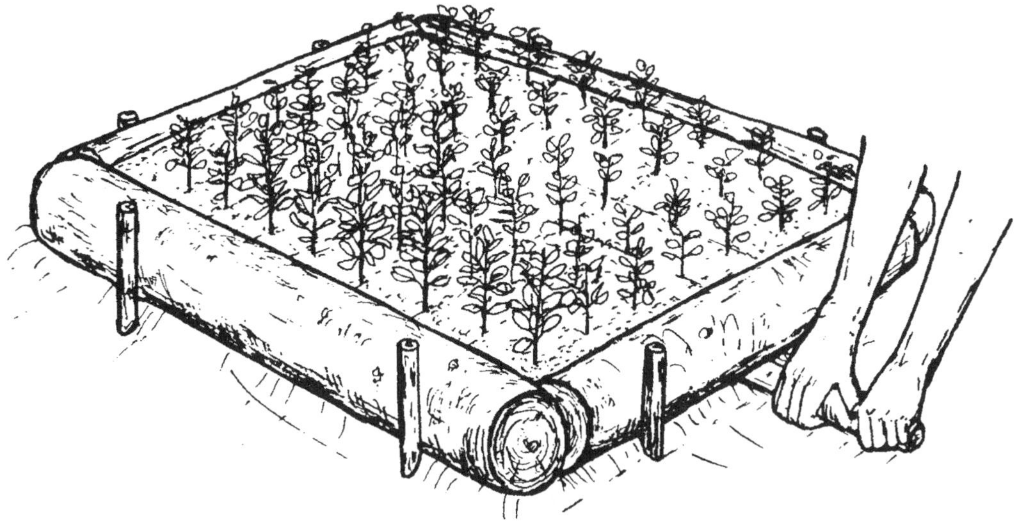

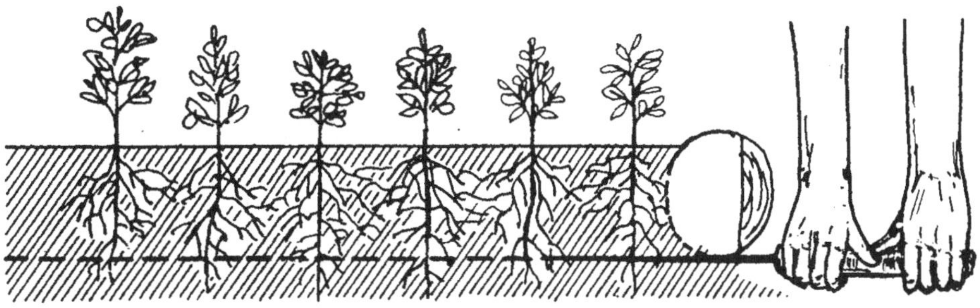

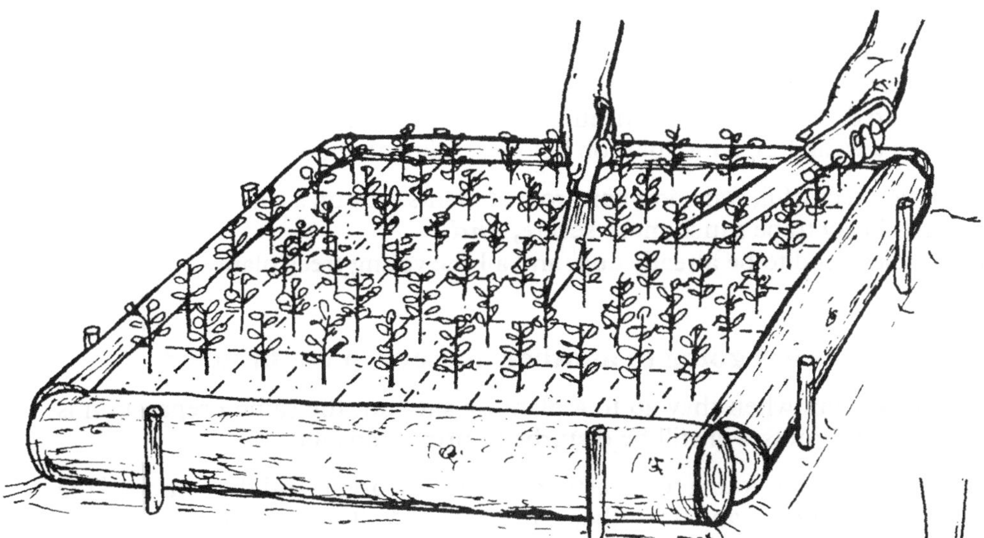

Regular cutting of the longer roots promotes good bushy root systems in the 'root-pruned soil block' method of growing young trees.

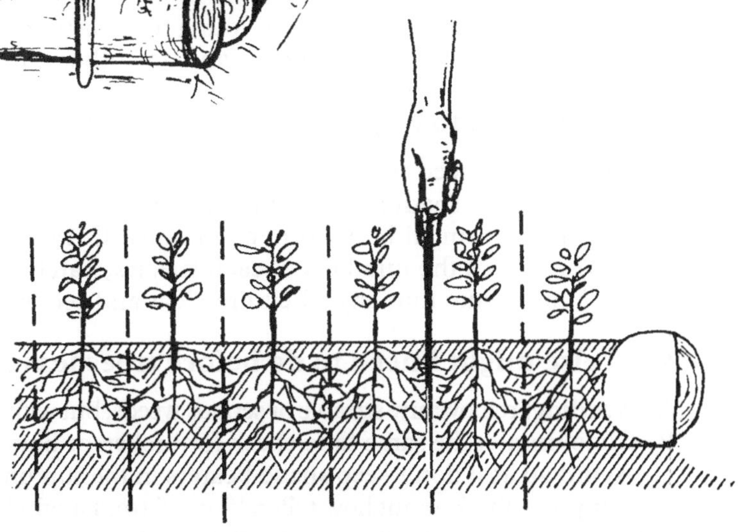

How should I reduce the amount of foliage, if this is needed?

In general, it is much better to reduce the size of leaves than to take them right off. Shortening them, by cutting (or even tearing) off half to three-quarters of the tip part:

(1) reduces water loss considerably; ***but***
(2) keeps the basal part of each leaf functioning properly;
(3) tends to discourage the premature outgrowth of side branches (C 12); ***and***
(4) is less likely to allow micro-organisms causing disease to enter (C 45).

For valuable research plants, you might consider keeping more foliage, and using a 'transplanting spray' which temporarily reduces the amount of water lost from the leaves.

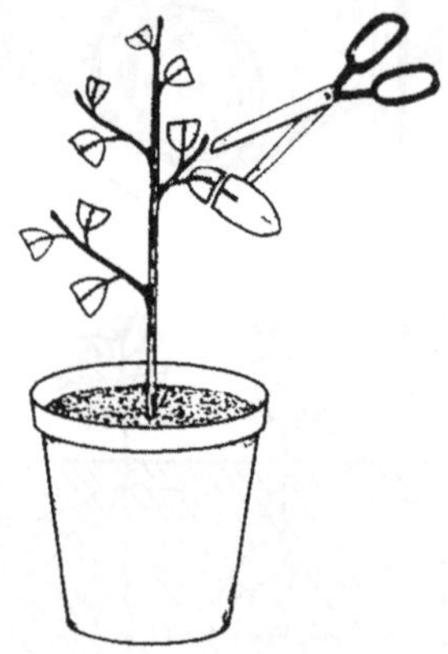

What about the transport of bare-root seedlings?

In cool climates, the soil is sometimes removed from the roots, because this greatly reduces the weight to be transported. This method is sometimes used in the humid tropics (for example for rooted cuttings of *robusta* coffee), but it needs to be done at night in order to reduce water loss, with the plants covered with polythene sheeting or moist leaves. To protect them from disease (C 45), drench in a weak fungicide solution before sending them out from the nursery.

Unless you know it can be successful, transport bare-rooted plants only on a trial basis, because:

(A) root damage is likely to be much more severe; ***and***
(B) the planting holes need to be already dug, and all the young trees planted before the sun is up next day.

Supposing the young trees are going to another nursery in containers?

There should be less stress than when they go to a planting site, but the same preparation is advisable because the soil and roots will still be disturbed during transport.

Don't forget to send a note with the young trees, listing the species, origin and the source of the seeds or cuttings, date of potting up, and any other relevant details (C 54, C 64-65).

And if they are to be planted in a field experiment?

In that case, you need to:

(1) alter the shading and watering regimes of the whole batch with extra care;
(2) keep them all in uniform conditions (C 48) while they are waiting to go;
(3) re-check the numbers of each species, batch or clone; ***and***
(4) make sure that any labels are clear and securely attached.

What else might I need to do before trees leave the nursery?

(A) **Weeding:** remove any sizeable or serious weeds, but with as little disturbance as possible to the soil;
(B) **Checking:** at the same time check for broken or burst containers;
(C) **Avoid stress:** don't keep shade-bearers in the sun unnecessarily;
(D) **Support:** think about how taller, 'floppy' trees might be protected in a carrying tray or vehicle from falling over, sliding about and getting broken, or rubbing against ropes or parts of the vehicle.

Protecting growing trees C 48
- *special conditions for research trees*

Why do young trees for research need special attention?

Because:

(a) their general environment should be as uniform as possible, to reduce unnecessary variation (C 7, C 69-N);
(b) the risk of various kinds of stress (C 41) needs to be kept to a minimum, so that damage and interruptions to growth are less likely;
(c) attack by pests or diseases (C 45), and damage from larger animals, wind or fire (C 3, C 40, C 46) all need to be avoided.

Are more uniform conditions needed just during an experiment?

That is when the care and checks described in sheets C 40-47 need to be especially well covered. But they are also needed when growing batches of trees to use later on in experiments (C 7), or for non-experimental research.

What is particularly important about the general conditions?

(A) *Light and shade.* The total amount of light that reaches each tree is likely to influence its growth:

(1) **directly** through the amount of sugars produced in photosynthesis (C 12, C 14); ***and***

(2) **indirectly** by affecting for instance:

(a) leaf and root temperatures (D 11 in Manual 4); ***and***

(b) the amount of water lost (C 13).

(B) *Moisture.* The drying power of the air also affects how much water is lost by each tree, while the availability of water in the soil is crucial to how quickly it can be replaced.
If severe water stress develops within a tree, hormones usually reduce its growth by:

(1) the closing of many guard cells (C 12) for several days, checking further water loss but also restricting the entry of carbon dioxide and so the formation of sugars; ***and***

(2) a slowing down of cell division and other processes (C 10, C 14).

(C) *Nutrients.* The amounts and availability of important nutrients in the soil of containers or beds can also influence growth considerably (C 14, C 33-34).

But won't the research trees differ from each other, anyway?

Yes they will, because of differences in:

(1) genetic constitution, unless they are all of one clone (C 5; and A 11 in Manual 1);

(2) previous environments; ***and***

(3) size, leafiness, extent of root system, inoculation with micro-organisms, and so on.

The aim is to have a set of research plants where all these sources of variability are kept as small as possible, so that effects of experimental treatments and other differences are more likely to show up (C 55, C 69).

How can I let each plant get roughly the same amount of light?

(a) By choosing a place in the nursery which is fairly evenly lit throughout the day (not too close to buildings, hedges or trees);
(b) By arranging the batch of research trees in an appropriate way within that area, spacing them out if necessary to avoid taller plants shading shorter ones;
(c) If shade is provided, by using a uniform material such as shadecloth, and extending it well beyond the batch of trees; ***and***
(d) By putting rows of other plants around the research trees to avoid 'edge effects'.

For some detailed research (C 62-A), young trees are grown (continuously, or from about 17.00 to 7.00 hours) in growth chambers where the environment can be closely controlled.

Should I water all the trees in the same way?

Generally yes, in order to keep the water balance (C 13) similar in all of them; ***but***
Sometimes no, for instance when some trees have a much greater leaf area than others, because of differences in initial size, genetic variation or experimental treatment.

Here a **conditional** method of watering (C 43) is needed, with some pots getting more frequent watering than others, which will help to keep the water balance similar.

What is the best way of applying nutrients evenly?

Depending on the circumstances, you could:

(1) Thoroughly mix in some rich soil, or a slow-release fertiliser, before potting or transplanting (C 6); ***or***

(2) Water the plants evenly once every 1-3 weeks with a **dilute** liquid feed, obtained from organic wastes or a dissolved, balanced fertiliser (C 33); ***or***

(3) Sprinkle small quantities of a solid fertiliser evenly on the soil around but not touching the trees, using the same weight or volume for each container or m^2 of bed.

Can I do anything about air humidity?

This can be increased to some extent around the plants by:

(a) choosing a sheltered site for the nursery (C 25), and protecting the research plants from the wind by planting shelter-belts and hedges (C 46); ***and***

(b) watering the ground around the young trees during dry weather.

Greater control over air humidity can be achieved by growing the plants in a greenhouse.

What's the point of having greenhouses in the tropics?

They can have several advantages when growing plants for research (C 7), and doing experiments with potted plants (C 15), through having:

(A) **more control over the growing environment,** for example:

(1) higher air humidity, except during periods of bright sunshine when even a shaded greenhouse will need to be well-ventilated to prevent temperatures becoming too high;

(2) preventing temperatures from dropping below a given value (C 41);

(3) having more uniform conditions, repeatable in each of a series of experiments;

(B) **protection from wind,** which could otherwise:

(1) mechanically damage or break leaves and stems (C 3);

(2) tend to loosen root systems and blow containers over;

(3) increase the risk of water stress within the young trees (C 41);

(C) **protection from rain,** for instance when:

(1) doing research on soil moisture or nutrients;

(2) making detailed measurements;

(3) using equipment that needs to be kept dry;

(D) **protection from animals,** such as:

(1) insects, particularly those that damage trees (C 45);

(2) larger animals that might eat or knock over research plants, or damage equipment (C 25, C 46); ***and***

(3) unwanted pollinators when making genetic crosses (C 5 and Manual 2).

Isn't the temperature hot enough anyway?

Not necessarily. Cool temperatures can occur, for instance:

(a) in the night, and sometimes on cloudy days;
(b) during periods when cooler air is drawn into tropical weather systems; ***or***
(c) at higher elevations, and on slopes near to mountains.

The optimum temperature for rapid shoot growth of tropical trees is often quite high. For example in young *Ceiba pentandra* it was found to be around 36°C, in the night as well as during the day. At 15°C, many species grow poorly, and at about 10°C most are damaged (C 41; and D 11 in Manual 4).

How can a greenhouse be heated?

(A) *By capturing the sun's heat,* and regulating the amount of ventilation according to cloudiness and the outside air temperature; ***and***
(B) *By using supplementary heaters,* when (A) is insufficient.
An experienced greenhouse attendant can adjust (A) and (B) manually, using appropriate heaters as needed.

If sufficient funds and a reliable electricity supply are available, the most accurate control system uses heaters and motors that open vents, both controlled by thermostats.

Waste of energy can be avoided by:

(a) avoiding draughts at night (for example under the door);
(b) using a time-switch to isolate the heating when it will not be required; ***and***
(c) including a switch that automatically cuts heat off when the vents are open.

Aren't greenhouses very expensive?

The cost of importing a large aluminium-frame greenhouse plus the sheets of glass from abroad could be quite considerable. However, polythene tunnels and their curved metal supports are much less expensive, and could easily be produced locally.

An even cheaper option is to make a 'home-made' greenhouse, as used for instance by families growing vegetables in Peru. This could be made out of curved plastic piping supported by wooden poles and covered with pieces of polythene sheeting stapled together (see A 31 in Manual 1). Another alternative is to make a 'lath house' with split bamboo roofing that gives some protection from the weather.

What about running costs?

Unless electricity is used, these are unlikely to add greatly to the running costs of the nursery, while savings can be expected through:

(1) better survival and growth of the young trees; ***and***
(2) advantages resulting from more precise research.

Some extra running costs could result from:

(a) maintenance of the greenhouse and its fittings; ***and***
(b) allowing for the *depreciation* of its value, in the expectation that replacement might be needed, say after 3-15 years.

Is it difficult to build a greenhouse?

Not particularly. Some useful hints are:

(1) choose a site that is nearly flat, but not liable to get waterlogged (C 20);
(2) avoid very heavy, poorly drained soils (C 23);
(3) sink some of the supports well into the ground, and secure them with concrete or stones hammered down, so that the greenhouse won't fall down or be blown away;
(4) face the door away from the prevailing wind direction, and take polythene sheeting under the soil level all around, to stop it being blown loose;
(5) make a well-fitting but easily movable door, to be opened when ventilation is needed; and consider the need for vents that can be opened and closed;
(6) check the air temperature inside on a hot day before putting in valuable trees.

Does it need any drainage?

Yes, this needs to be considered before building the greenhouse, or its soil may become too wet. Remember to think about the water from both:

(A) *rain outside,* which could be collected for use in a large drum, and the overflow carried away; ***and***

(B) *the excess from watering inside,* which might be dealt with by a slightly sloping 'blind' drain (C 23), covered with smaller stones or coarse gravel. This could run down the middle of the greenhouse, serving also as a path, and then provide water to a drier part of the nursery.

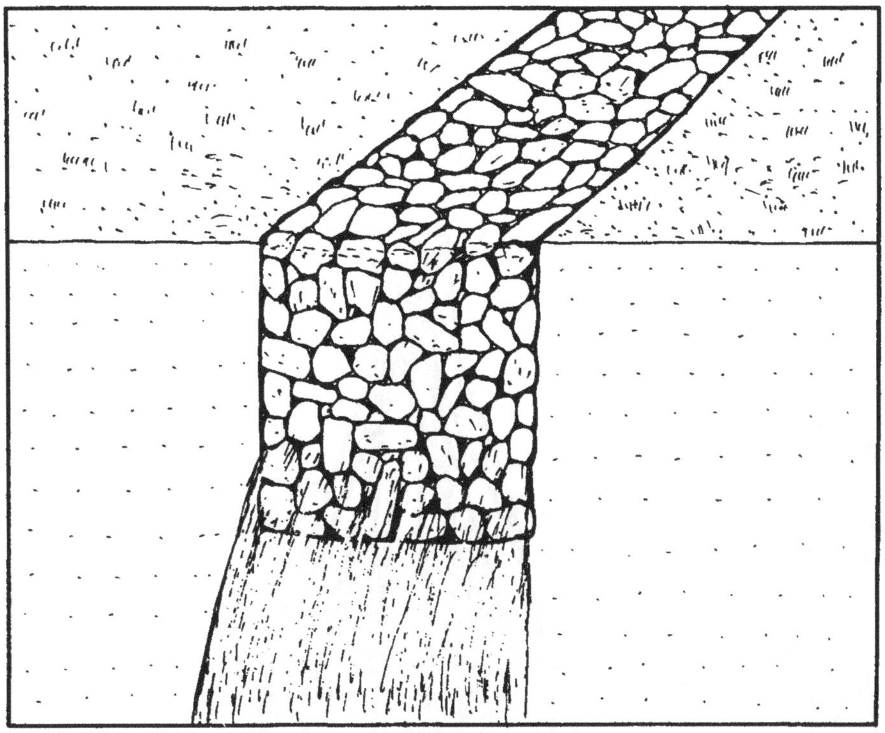

Blind drain.

Should I put good soil on the floor of the greenhouse?

Yes, when plants are to be sown and/or planted directly into it; ***but***
No, if trays are going to be used for germinating seeds (Manual 2), and containers (C 6) for growing on the young trees, where you could use ordinary soil, gravel or concrete slabs on the floor.

A convenient alternative for experiments is to put the containers on *staging* or old tables that raise them 0.75-1.0 m above the ground.

Will the young trees in the greenhouse need more frequent watering?

Often they may, because they will generally not get any from the rain, and the temperature inside may be higher, so that they may lose more water; ***but***
Sometimes they may not, for instance in windy, dull weather.

Do greenhouses need a lot of maintenance?

Not a great deal, provided that they have been carefully erected. Points to check on (C 40) include:

(a) mending any tears in the polythene at once, and replacing broken panes of glass;
(b) periodically firming-in any loose supports, and replacing those that have become weak;
(c) avoiding any accumulation of dead leaves and waste materials inside or nearby;
(d) repainting non-durable wood in dry weather, and watching out for termite attack.

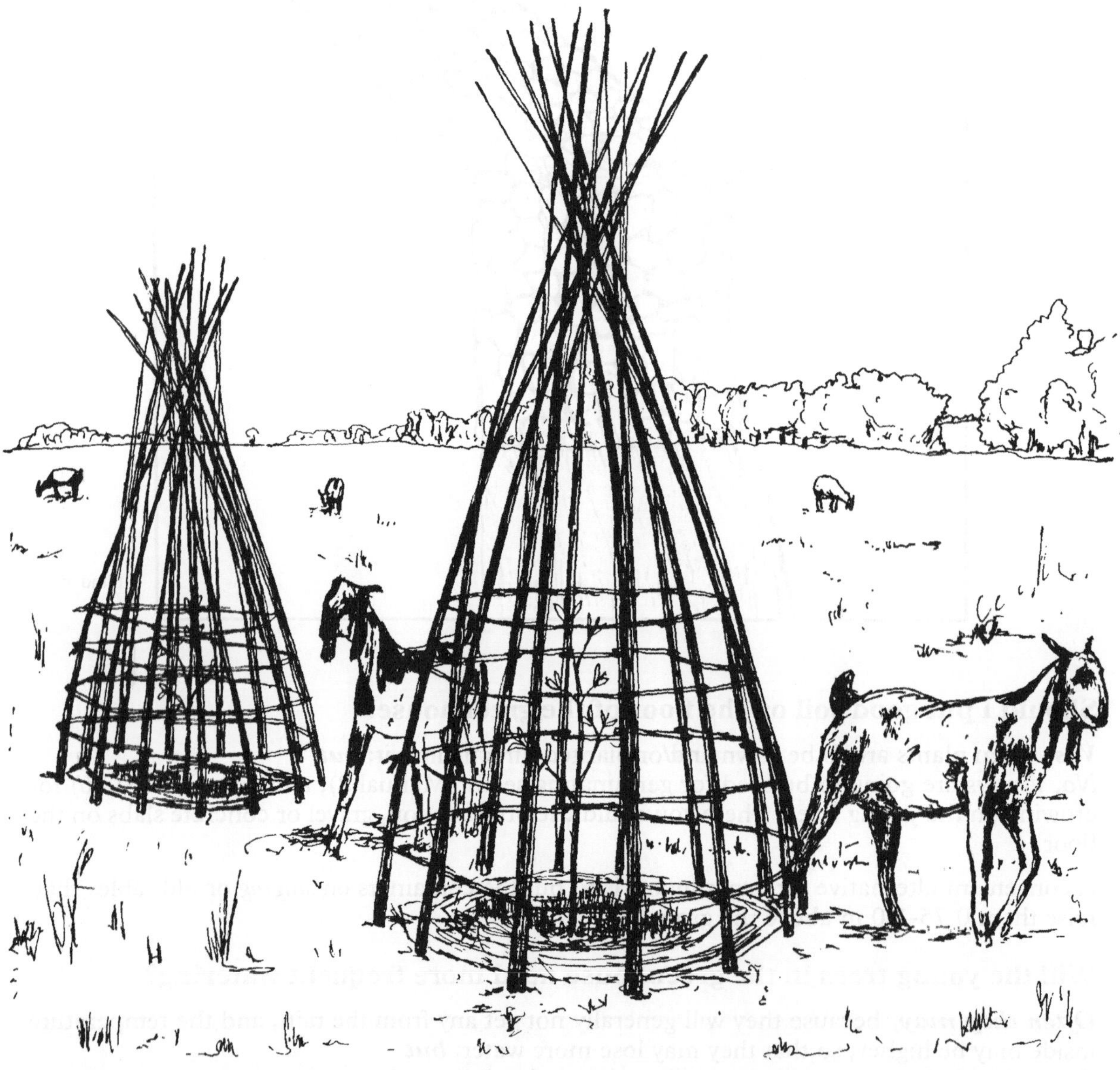

RUNNING A TREE NURSERY

Running a tree nursery C 50
- introduction: training and maintenance

What sort of things are involved in running a tree nursery?

Being able to keep in mind a whole lot of different points, including:

(1) the scientific basis of growing good young trees for planting (C 10-15, C 30-32);
(2) technical aspects of good nursery practice (C 20-26, C 33-34, C 40-48);
(3) questions of economics: avoiding waste, spending wisely and meeting objectives;
(4) encouraging the people involved to keep learning more about the job (C 52);
(5) planning ahead and keeping good records (C 54); ***as well as***
(6) getting everything ready for the day-to-day work of the nursery (C 51).

Which are the most important things?

Learning about the job yourself, and building a good team of nursery staff and workers.

Why is that the top priority?

Because:

(A) there is always more one can learn about tree nurseries;
(B) remembering the many important points about growing good plants will be harder if time and energy are wasted through lack of co-operation and in settling disputes; ***and***
(C) a good working atmosphere is the best way of making steady progress, and of dealing with any problems promptly (C 3).

So I need a lot of patience?

Yes; and as a 'key person' (D 4 in Manual 4) you will need not only an understanding of how to treat other people, but:

(a) a basic grasp of how to run an enterprise;
(b) dedication to the job and determination to overcome obstacles;
(c) the ability to handle several different things at once; ***and***
(d) a willingness to try out ideas to see whether they work in your situation.

Does the science really matter much to the practical grower?

Yes it does, because the general principles described in these manuals are:

(1) not an optional extra, but the first foundation for success; ***and***

(2) based on research on trees that has been done for many years all over the world.

For instance, making do with less suitable genetic origins (C 5) might mean poorer trees being used for a long time, or even that a valuable species was mistakenly ruled out.

Isn't growing trees too easy to need to bother about technical details?

Not at all. It may look easy, but local experience throughout the world suggests that:

(a) there are many potential pitfalls (C 3), any one of which might damage or kill the young trees, and so hamper the production of enough good planting stock; ***and***

(b) planning a tree nursery carefully (C 20-26) and learning how to look after young trees in it (C 40-48) can save a great deal of time, effort and money.

A lot of useful trees, shrubs and climbers used to come up naturally, but now may not do so (C 1-2). Specific technical information for growing these is often lacking.

Will these things help to make an economic success of the tree nursery?

Such technical points form the second foundation for nursery success. Next comes the third aspect—important questions of economics, including costs and funding, but also:

(1) planning a logical and efficient internal layout for the nursery (C 22);

(2) deciding for each tree species whether to root cuttings (Manual 1), sow seeds (Manual 2), collect wildings (C 2) or use more than one method of propagation;

(3) working out how many young plants will be needed each year, and how much nursery space that will require (C 63-A,B);

(4) obtaining adequate supplies of materials (C 24) and tools (C 51) in good time, so that nursery work is not held up;

(5) maintaining tools, any equipment and vehicles in good condition by regular cleaning, checking, servicing and repairs;

(6) thinking about potential accidents or breakdowns (C 3), and making sure that daily and weekly checks are done (C 40, C 66); ***and***

(7) discovering the simplest and quickest way of doing individual tasks, or successfully combining two.

Good economics also takes into account the costs of ***not*** doing something, and the intrinsic value of a well-trained team (C 52).

But supposing there is a shortage of money for the tree nursery?

There usually is! But:

(a) small tree nurseries at home or on the farm (C 20-22) can usually function without much money; ***and***

(b) a lot can be done in moderate sized nurseries by using materials and containers that are cheap, or free because otherwise they would be wasted (C 33).

Large nurseries are the most likely to suffer if staff and workers are not paid for months.

How important is training?

It is vital to success, and can be seen both as:

(A) **training on the job,** consisting of such points as:

(1) explaining to staff and workers why careful handling of young trees (C 40-42) is important; and correcting any bad practice at the time it occurs;

(2) demonstrating how the regular watering is to be done (C 43), and making sure that the methods are learnt and followed;

(3) showing people what is needed when another job is to be done, and explaining why it should be done that way (C 52); ***and***

(B) **special courses,** where a new approach needs to be learnt, such as:

(1) changing from a temporary to a permanent nursery (C 21);

(2) rooting cuttings of trees (C 5; and Manual 1); ***or***

(3) inoculating young nursery trees with micro-organisms (C 31-32).

What about planning ahead and record-keeping?

These go together, helping the work to run smoothly (C 54). Some examples are:

(1) when you have learnt from experience the best time to set cuttings and sow seed, then:

(a) the planting stock is likely to be the right size (C 34) at the time it is needed;

(b) you won't be wasting effort and materials keeping it longer than necessary in the nursery, or find yourself trying to stimulate growth with fertilisers (C 33); ***and***

(2) if you make sure that simple basic records are properly kept (C 64-66), you can easily look up the data for previous years instead of having to guess.

And the day-to-day running?

This is really a question of organising a piece of the overall job, involving:

(A) routine operations like watering (C 43), weeding (C 44) and checks (C 40); ***and***

(B) specific tasks for the day.

Omit (B) rather than (A) if there is a sudden urgent request, or an emergency.

Does work with experimental trees take longer?

Yes, it usually does, though some tasks may be the responsibility of the research worker. More time will be needed in order to:

(a) understand what is involved (C 7);
(b) produce the necessary uniform growing conditions (C 15); ***and***
(c) look after the young trees particularly carefully (C 48).

Periodic observations (C 55), measurements and analyses (C 67-69) may also be needed.

Does maintenance matter much in a tree nursery?

Yes it does, because 'a stitch in time saves nine'. Maintenance includes not only checking, servicing and repairing any vehicles, buildings and equipment, but also:

(1) cleaning out filters and doing timely repairs to the nursery water supplies (C 24);
(2) looking after tools (C 51); ***and***
(3) keeping up fences, gates and hedges (C 46).

It is also important to 'maintain' and occasionally review the arrangements for responding to emergencies such as flooding, fires (C 3) or snake-bites (D 66 in Manual 4).

Running a tree nursery C 51
- *tools and materials for the nursery*

Which sort of tools are required for nursery work?

It is possible to run a small nursery quite effectively with just a few basic tools, so a lack of money need not prevent good trees being grown. For instance, you could use:

(1) the farm tools used locally for cultivating the soil and for cutting poles and weeds;
(2) a shovel for mixing up potting soils (C 6), and a hand tool for filling the containers;
(3) a pail and an old plastic bottle or can with holes in the bottom for watering (C 43);
(4) convenient pieces of wood or metal to lift out newly rooted cuttings and small seedlings (Manuals 1 and 2); for potting up or transplanting (C 42); and for weeding (C 44).

What other tools would be useful if some funds were available?

(a) **a spade** for digging out materials for potting mixtures (C 6), and for preparing beds and level standing ground for young trees in containers;
(b) **a sieve** for removing stones and other large items from soils and composts (C 33);
(c) **a funnel** for filling or part-filling pots (C 42);
(d) **a pickaxe** or **crowbar** for fencing (C 46) and making access paths (C 22);
(e) **a hammer and nails, a saw** and **a ladder** for such jobs as making wooden seed-boxes, fencing and general improvements and maintenance around the nursery.

Do such tools need any maintenance?

Some may need sharpening, an occasional replacement handle and perhaps hammering straight. All will last longer if they are:

(A) used with reasonable care; ***and***
(B) cleaned, dried and kept inside afterwards.

Which kinds of materials may be needed?

The first requirement is for continuing supplies of suitable components (C 24) for making rooting media (A 35 in Manual 1), seed germination media (Manual 2), potting mixtures (C 6, C 63-D), and for improving nursery soils (C 23).

Other needs depend on whether the nursery is temporary or permanent (C 21), and also on its size (C 22).

What would be the minimum for a temporary nursery?

Besides soil mixes, you might need a supply of:

(1) free or waste items that could be used as containers (C 6, C 33);
(2) suitable materials to use in putting up shade (C 41); ***and***
(3) a small notebook and pencil for plans and records (C 54).

And what else if the nursery is likely to be permanent?

Amongst other items, you might also need:

(a) larger supplies of several kinds of containers;
(b) plastic sheeting and wood for making poly-propagators (A 31 in Manual 1);
(c) hand-sprayers for keeping high humidity (A 34), restoring wilted plants (C 13), and controlling diseases and pests (C 45);
(d) posts, gates and fencing materials (C 46);
(e) broken stone or coarse gravel for improving access roads and paths (C 22);
(f) foundations, wood, roofing and preservative for building a shed; ***and***
(g) sundry items like labels (C 54) and perhaps fertilisers (C 33).

Sieving larger quantities.

What other things might a large nursery require?

You could also consider whether to:

(A) put up a wooden framework and shadecloth over part or all of the nursery;
(B) install a better water supply (C 24);
(C) buy a small portable electric generator (or a *micro-hydro* unit) for lights and power tools; ***or***
(D) get a four-wheel trolley, van or pick-up truck for transporting materials and young trees.

Running a tree nursery - *the nursery team* C 52

Who needs to be part of the team?

Everyone who is involved with the work of the tree nursery. For example:

(A) all members of the family, for a small nursery at home or on the farm;
(B) two or three working colleagues, for a medium size, jointly-run nursery;
(C) all the staff and workers, for a big nursery; ***and additionally***
(D) specific people at headquarters and abroad, for an international research project.

How can a good team spirit be built up?

The way the 'key person' (C 50) works is the crucial point. It is important to:

(1) explain clearly what is involved in a successful tree nursery;
(2) show just how each job is to be done;
(3) decide clearly who is going to do what when; ***and***
(4) make sure that the work is done promptly and properly.

In larger nurseries, it might also be useful to:

(5) hold regular staff meetings; ***and***
(6) ask individual staff members to be specially responsible for different parts of the nursery work.

Supposing someone is causing problems?

What to do depends upon the circumstances. If for instance a worker or a staff member often comes late, makes mistakes or is slow to learn, you could explore whether he/she:

(a) had previously been used to a different style of working;
(b) hadn't understood properly what was required;
(c) has been having a difficult time, because of bereavement, illness or other reasons;
(d) is in fact beginning to improve.

However, if someone:

(1) cannot be relied upon;
(2) keeps on saying they know best;
(3) upsets other team members by arguing or abuse; ***and***
(4) does not show signs of improvement when warned more than once; ***then***

It is likely that the team would be better off without them.

Aren't staff meetings just a waste of time?

Sometimes they could be, for example:

(a) if no preparation has been done or agenda drawn up;
(b) if the discussion is ill-disciplined, contentious or long-winded; ***or***
(c) at times when there are far more important priorities.

However, provided these pitfalls are avoided, they can often be useful in:

(1) encouraging good communication amongst the staff and workers;
(2) listening to views on problems and anticipating difficulties;
(3) promoting the making of clear, appropriate decisions, and planning ahead (C 54);
(4) allowing easy checking of whether things were done to plan, or why they were not.

But aren't decisions up to the person in charge?

In the long run they may be, but:

(A) sensible decisions depend on first being well-informed; ***and***
(B) people work better if their ideas, experience and grumbles are taken into account.

What about training?

This is very important, for both staff and workers (C 50).
In one sense, training may go on for much of the time, as people continue to gain further experience of different aspects of nursery work, changing seasons and various species.
As well as this, they may also need to gain new knowledge about fresh approaches or improved techniques, either on the job or by attending a specific course.

Will training increase the output of the nursery?

Yes, it can decrease the losses of young trees; and also improve nursery efficiency by:

(1) increasing the quality and uniformity of planting stock;
(2) reducing the time the young trees spend in the nursery; and damage to them;
(3) avoiding waste of water and materials;
(4) decreasing the time taken to do nursery jobs, and the amount of supervision needed; ***and so***
(5) keeping down the running costs.

Can nursery maintenance be improved by training?

Yes, especially when:

(a) a maintenance schedule is drawn up for all the appropriate aspects (C 50); ***and***
(b) the checks, servicing and any repairs needed are actually carried out.

Isn't all this unnecessary for a family nursery?

Not really, for many of the above points can be adapted to these circumstances.
A system where only one person knows how and when to do the work is liable to break down when they are busy, away from home or taken ill.

How about a community nursery?

Similar questions apply here, and it is especially important to achieve:

(1) good communication between all those with an interest in the tree nursery and the planting of trees (C 53; and D 5 in Manual 4);
(2) fair sharing out of the work, the available information and the young trees; ***and***
(3) effective checking that the jobs are being done, and control over standards of work.

Otherwise, despite people's efforts, it is unlikely that successful tree plantings will result.

Running a tree nursery - *useful contacts* C 53

Why would I need to get in touch with other people?

Because there can be a lot of mutual benefits from not just working away on your own. It could be helpful to link up with people and organisations:

(A) in the neighbourhood;
(B) elsewhere in your region or country; ***and***
(C) abroad.

What sort of links could I make nearby?

For example you might be able to:

(a) learn more about growing good trees from experienced individuals, and from any local demonstrations, exhibitions or workshops;
(b) use your own experience to help a new project get off to a good start;
(c) exchange ideas, plants, cuttings or seeds with other tree nurseries and growers; ***or***
(d) sell young trees to people or groups wanting to plant them.

Who might give regional or national advice?

Valuable ideas and suggestions can sometimes be obtained from:

(1) government forestry, agriculture and parks departments, extension offices and research stations;
(2) private growers, large and small; ***and***
(3) agroforestry projects run by local non-governmental organisations.

Remember that it is not a one-way process - your views and knowledge are important too!

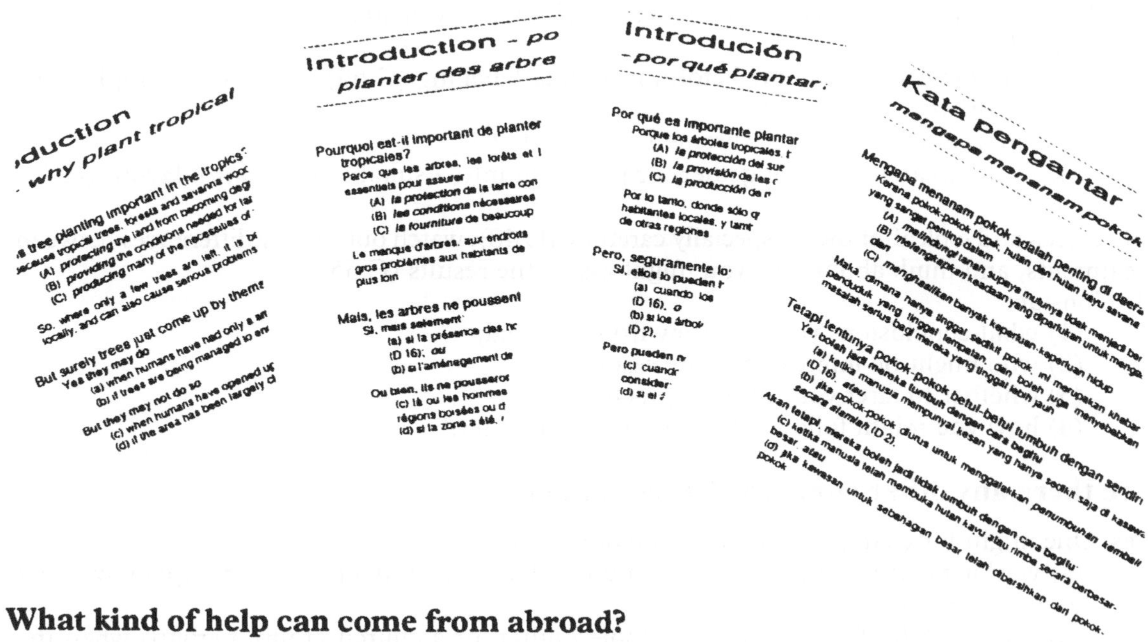

What kind of help can come from abroad?

(a) Publications, practical manuals and videos about nursery techniques and tree species (C 61; and D 70-72 in Manual 4);
(b) Data about which genetic origins of commonly planted trees have been found to thrive in different climates and sites (C 62-E); ***and***
(c) Findings of international projects concerned with growing trees, especially those that involve joint studies with local farmers or communities.

Will all this be relevant to me?

Probably not; but it is well worth checking up on what is available, because:
(1) some of the possibilities might be just what you need;
(2) the benefits could be considerable and long-lasting; ***and***
(3) it may cost you little or nothing.

Progress in science and technology can often be stimulated by the exchange of ideas and experience among people working in widely different fields.

But it could be difficult to find out about such things!

That's often so, which means that everyone is the poorer. Here are a few hints:
(a) keep your eyes open for newly planted trees that are thriving, and go and enquire;
(b) look out for other nurseries, whatever species are being grown;
(c) ask your friends to tell you if they hear of anything relevant; ***and***
(d) listen to what the children have been learning about and doing at school.

Supposing things are being stolen from the nursery?

As well as 'negative' action like putting up a strong fence (C 46), having a night watchman or notifying the police, you could try more positive steps such as:
(1) contacting local mayors, chiefs, elders or other people for advice;
(2) explaining the importance of tree planting to teachers or to a community meeting;
(3) giving local people a stake in the success of the nursery by employing some of them; ***or***
(4) advertising how cheap the trees are to buy, rather than steal.

What if I want help with an informal trial?

(A) Look through the points made in sheets C 7 and C 15, and the relevant parts of D 6 and D 55 in Manual 4;
(B) Grow a batch of uniform plants that is larger than you need, giving you a chance to select the ones for the experiment;
(C) Read and think about the particular aspect you are interested in studying, and discuss it with anyone who has relevant experience;
(D) Work out the treatments you want to apply, how many plants you will need for each, and how you will avoid bias;
(E) Get everything ready the day before, including labelling (C 54), and try to complete the setting up of the trial in one day;
(F) Make appropriate initial measurements (C 55, C 67);
(G) Mix all the treated plants together, or put them into a suitable *randomised* layout (C 15, C 62-F);
(H) Check and look after them especially carefully (C 48), watch out for any differences between treatments, and think about how you might assess the results (C 55).
(I) Consider:
(1) what conclusions you can draw from the trial;
(2) who might be interested in the results;
(3) whether to alter your nursery practice; ***and***
(4) how you might follow up this trial with another.

Are there any other contacts that could help?

Yes. One might for example explore the possibility of:
(a) working closely with a local community, farmers' co-operative or company wanting to plant trees; ***or***
(b) giving a talk, for instance to school-children or a church group, perhaps when the government is planning a tree-planting week.

Running a tree nursery C 54
- *planning ahead, labels and records*

Is planning ahead particularly important for a tree nursery?

It is crucial to success.

Why is that?

Because with forethought one can have:

(1) fewer breakdowns, and reduced damage to and loss of trees;
(2) less waste of water, materials and effort; ***and***
(3) more thriving young trees, produced at a lower cost.

What kind of things need to be thought about?

(A) **Nursery layout:** how the areas within the nursery for beds and containers, working space, storage of materials, and so on, can be arranged for greater convenience and efficiency (C 22);
(B) **Risks:** reducing both the likelihood and the extent of any damage to young trees caused by accidents (C 3), various types of stress (C 41), or diseases and pests (C 45);
(C) **Smooth running:** building up a good nursery team (C 52), doing regular checks (C 40, C 66) and keeping sufficient tools and materials (C 51), so that the work is not held up by lack of understanding, poor communication or shortage of an essential item;
(D) **Adaptability:** having alternative work thought out, in case weather conditions, illness or delays prevent the planned job being done;
(E) **Costs:** Estimating the approximate overall costs of the various nursery operations and purchases, avoiding waste, and spending wisely to achieve the production of good planting stock (C 50); ***and***
(F) **The future:** including changing needs for planting stock; using experience to avoid problems; seeing a need for extra contacts (C 53); and planning improved schedules for the next season.

Is labelling really required in a tree nursery?

It is important whenever confusion could arise, for example between:

(1) various batches of seeds and seedlings of the same or similar tree species;

(2) bags containing shoots for vegetative propagation and rooted cuttings of different clones;

(3) plants for research, especially before treatments are applied and during the experiment.

How many labels will I need?

Two labels are a minimum, for instance with:

(a) one inside a bag of seeds, and one on the fastening tie;

(b) one at the start and one at the end of a batch of seeds that have been sown; ***and***

(c) one in the first and one in the last of a set of pots, or in a section of nursery bed.

Labelling each plant is sometimes needed, for example with:

(A) **vegetative propagation,** in order to gain the benefits of knowing which clone a tree belongs to (C 5; and A 11 in Manual 1); ***and***

(B) **experiments,** so that the influence of treatments and of genetic origin upon the growth of individual trees can be studied (C 7, C 15, C 55 and C 69).

Which kind of labels are most suitable?

(a) **Material:** labels are often made of plastic, metal or wood.

(b) **Type:** those that have a blunt point at one end and a small hole near the other are convenient, as they can be pushed into the soil, and later attached to the tree with a tie. If holes are missing, they can be drilled through a number of labels held tightly together.

(c) **Size:** relatively small labels are usually most suited to nursery work, for instance about 8-12 cm long, 1-2 cm wide and 0.5 -2.5 mm thick.

(d) **Colour:** writing is more distinct on white or pale coloured labels, so it may be worth painting wooden or metal labels white; and the paint might also make them last longer. Plastic labels come in many colours, which can prove useful for temporary labelling of young trees that are to receive a particular treatment.

Can I make my own labels?

Yes, they could be made for instance out of:

(1) thin, smooth pieces of wood, bamboo or waste veneer;

(2) plastic containers or strong bags, cut up into strips; ***or***

(3) tough leaves that rot only slowly.

If the young trees are in containers, an alternative might be to label the pots.

What should I write with?

(A) **Pencil** is often the best with plastic and metal labels. Press firmly with a good HB, H or B grade pencil;

(B) **Waterproof felt-tip pens** can be effective on wooden labels, and also directly on to clean, dry stems of trees with smooth bark, and the outsides of containers. Note that 'waterproof' ink may wash off or fade on some plastic and metal surfaces during propagation.

Do I need to write a lot on the labels?

Only the essential points, in an abbreviated form (see A 64 in Manual 1), provided that the full details can easily be found in the records.

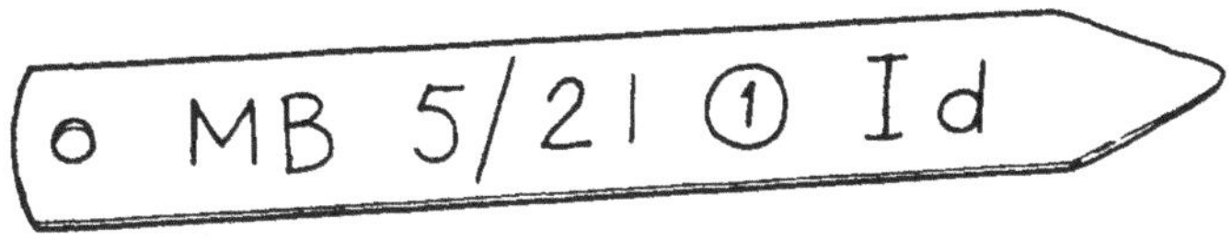

How should I attach the labels?

(a) *For small plants:* push the label into the soil, avoiding damage to the roots;
(b) *For larger plants:* use a loop of wire, strong string or a specially manufactured tie to attach the label to the tree. Do this loosely enough for the stem to grow thicker without becoming 'strangled', but not so large that the loop comes off.

Some types of thin plastic labels are made to be looped around the tree and passed through a slot. However, they can be difficult to attach if leaves or branches are crowded together, and they tend to become brittle after a while.

What other labelling problems might I have?

It is quite difficult to provide permanent identification of young trees in the tropics, because:

(1) the writing may become indistinct, or disappear altogether;

(2) labels may become brittle, tarnish or rot;

(3) they may get blown away, buried in the soil or chewed by certain animals;

(4) labels can be misplaced when transplanting or potting up (C 42) and during transport (C 47). It is best if they are routinely transferred from the soil to the plant, and from the main stem to a suitable side branch, as the trees get bigger; and are rewritten or replaced as needed.

Won't records prove to be a waste of time?

Occasionally they could do, if, for instance:

(1) only a few trees are being grown; ***or***

(2) the people involved have long experience with the kinds of trees being grown; ***but***

Generally they are definitely useful, in order to:

(a) learn from experience, so that plans are steadily updated and losses minimised;

(b) know the genetic origins of trees (C 5), so that those that are proving more successful can be used again, while those less suited to the district can be replaced by new selections (C 5); ***and***

(c) lay out experiments without bias (C 15), and know which trees received what treatment.

What are the key things to record?

Three of the most important points for tree nurseries are:

(1) **origins** and **dates** of collection and receipt of all the seeds, cuttings and plants coming into the nursery. A sample record sheet is shown in C 64;

(2) **records** of each batch of plants grown in or received by the nursery (C 65); ***and***

(3) **details of checks** made during propagation (C 66).

Other points might include times taken for jobs, cost of materials, and observations on the progress of the young trees (see also A 64-68 in Manual 1).

Are there some practical hints on keeping records?

(a) Write brief but clear notes on the same day, not leaving it till later;

(b) Photocopy a simple form for records and assessments (C 67) that will be repeated several times, and use a clipboard to make writing easier; ***and***

(c) Keep the records tidy in a looseleaf file or notebook.

Can record-keeping help with planning ahead?

Yes, this is one of its main advantages. For example:

(A) next time, the dates of propagation can be moved forward or back, in order to obtain planting stock of the desired size (C 34) at the best planting time;

(B) a close watch can be kept during stages when trees previously suffered from stress (C 41), or when losses occurred from pests or diseases (C 45); ***and***

(C) other improvements in technique can more easily be made.

Running a tree nursery C 55
- assessing pot experiments

What needs to be recorded during an experiment?

For most experiments one needs to make:

(A) **Observations** on the young trees, especially any differences that appear between those receiving a treatment and the controls, or between plants of various genetic origins; ***and***

(B) **Measurements,** generally repeated, about how fast the young trees are growing.

Sometimes these basic types of assessment might be supplemented by:

(C) **Destructive sampling,** in order to obtain estimates of *dry weight* or to study chemical composition;

(D) **Detailed study** of processes going on in the trees, such as manufacture of sugars (C 10), uptake of nutrients (C 14) or changes in water content (C 13); ***or***

(E) **Recording** features of the environment that the trees are growing in.

When should assessments be done?

This depends partly on:

(a) the type of experiment, and how long it is likely to last;

(b) what appears to be happening in your experiment, as well as any previous experience of what may be expected.

Useful assessment times are:

(1) at the end of the experiment;

(2) at the beginning;

(3) when changes seem to be occurring rapidly; ***and***

(4) at regular intervals during the course of the experiment.

NOTE: **in all cases, remember to write down the date of the assessment (C 67).**

Could most observations be left to the end of the experiment?

No, because it is usually much better to write a simple note on a record sheet or notebook, ***at the time that you first notice the change,*** because it may not be visible later on.

Soon afterwards, it may be useful to do a scoring assessment (C 68).

What sort of observations might I do?

Some examples are to record when you noticed:

(a) alterations in the **patterns of shoot growth;**

(b) **colour** changes; ***or***

(c) signs of **stress** or **damage**.

What sort of changed patterns of shoot growth might I look for?

(1) New shoots starting to expand after a dormant period (C 12);

(2) The end of a period of stem extension and leaf expansion;

(3) The loss of old leaves; ***or***

(4) New branches beginning to form.

Why not just measure these differences?

Measuring shows ***how fast*** growth is occurring, and may provide sufficient information about height growth when the leading shoots of the young trees grow continuously, producing new leaves and extending internodes without a break.

Observations can provide additional important data, especially on trees where the shoots grow by successive 'flushes' (C 12), such as *Khaya, Lovoa* and mango, and also about other changes in a tree's development of a ***'yes/no'*** type.

Recording when a new flush occurs could help to show up any:

(A) **different responses** of clones or varieties to the **same conditions;** as well as
(B) **effects of treatments** on trees of **similar genetic potential.**

Is it easy to see when a new flush starts?

The date of budbreak (C 12, C 68) can generally be pinpointed most accurately by:

(a) noticing expanding buds, or several small leaves emerging at the tips of shoots; ***combined with***
(b) finding that measured shoots which were previously not increasing have become longer.

What about the time that shoot growth stops?

The start of bud dormancy (C 12) is not so easy to spot, but can be detected by:

(1) no more new leaves appearing at the shoot tips;
(2) the leaf (or leaf pair) nearest to the tip being already fully grown and often becoming a darker green; ***and***
(3) a measured stem ceasing to increase in length.

In some tree species, the growing point at the tip of the leading shoot shrivels up at this stage, and further growth in height will be made from a lateral bud.

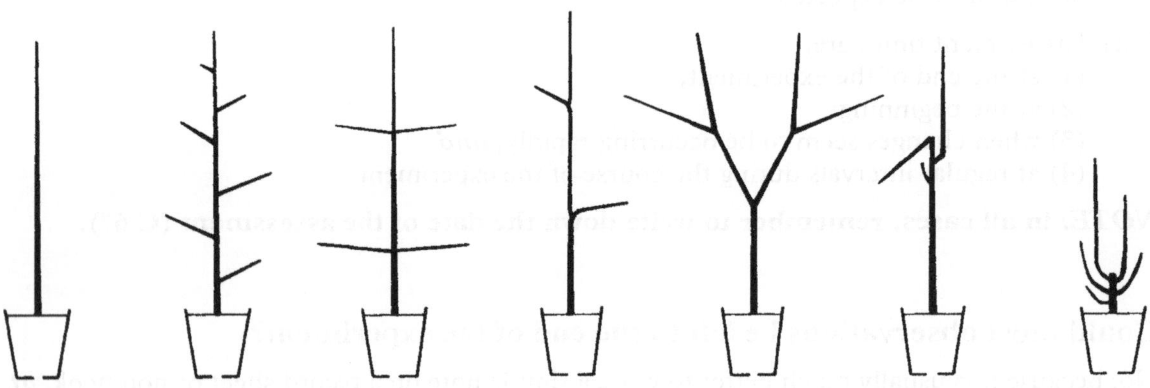

Common types of branching.

Where do new branches appear?

Branches may grow out from buds in many different places (C 12), according to the branching habit of the species, but they often come from:

(a) **newly formed lateral buds** near the tip of the main shoot, as it grows;
(b) **existing lateral buds,** often those near the tip of the *previous* flush;
(c) **older lateral buds,** especially when shoots have been slightly damaged; ***or***
(d) **small, inactive basal buds** or **newly-formed buds** in replacement shoots growing out from pruned plants or coppiced stock plants (A 21 in Manual 1).

Do branches need to be recorded?

A note of the date when branches first appeared can be quite important in comparing different species or genetic origins for 'branchiness' (A 13 in Manual 1), and might well help to distinguish the effects of treatments.

You could also make observations on the branching habit (C 68), and record whether all branches continue to grow obliquely, or some are turning up and becoming main stems.

How can one score colour?

(1) Look through all the trees in the experiment to see the range of colours present;
(2) Select 5-10 sample leaves showing a sequence of colours, for example from pale yellowish green through to dark green (preferably taking them from similar plants that are not part of the experiment);
(3) Attach these sample leaves to a piece of card, and number them in order;
(4) Do a quick trial run, matching leaves on experimental trees with those on the card, in order to get your eye in;
(5) Assess the whole experiment (C 68), giving a score to each tree, or to comparable sample leaves on it.

Couldn't photographs be used?

Yes, they can be valuable in recording what the colours were, and more generally for:

(a) recalling just what was occurring in an experiment; ***and***
(b) presenting the results to other people in discussions, talks or exhibitions.

Taking a picture of every young tree in a trial is seldom possible, but photographs of some of them in 2-4 lines, arranged in treatments or genetic origins, can be very informative.

What sort of damage needs to be recorded?

For example:

(1) **Loss of shoot tips,** especially of the leading shoot, because this may influence both height growth and branching habit. It can cause problems with measured shoots getting 'shorter', and difficulties with which replacement shoot to record; ***and***
(2) **Reduction of leaf area,** through leaves being partly or completely eaten, torn by the wind or distorted by virus attack (C 45), all of which could affect growth.

Try to distinguish at the time between damage that might be linked to treatment or genetic origins, and accidental or chance breakage.

How can damage be assessed?

(A) ***By counting*** the number of damaged:

(1) plants in each treatment or origin; ***or***
(2) leaves or shoot tips on each plant.

(B) ***By scoring*** the extent of damage (C 68). You could, for instance, assess whole plants or sample leaves, using a scale of 0 (no damage) to 5 or 10 (very damaged).

Is it really necessary to do a whole lot of measuring?

That depends a great deal upon the circumstances. In general, try and avoid either:

(a) ending up without important information about the results which could easily have been obtained; ***or***
(b) collecting a mass of figures for no particular purpose, or just because a previous (but no longer appropriate) decision had been made to do so.

What are the most useful measurements to make on trees?

The commonest simple, non-destructive ones are of the:

(1) **height** of the tree;
(2) **diameter** (or girth) of root-collars or stems;
(3) **numbers** of leaves that are produced or fall off;
(4) **length** of leaves or branches, and sometimes of individual internodes (C 12).

The ***gain*** that has been made since the experiment started is often a valuable calculation.

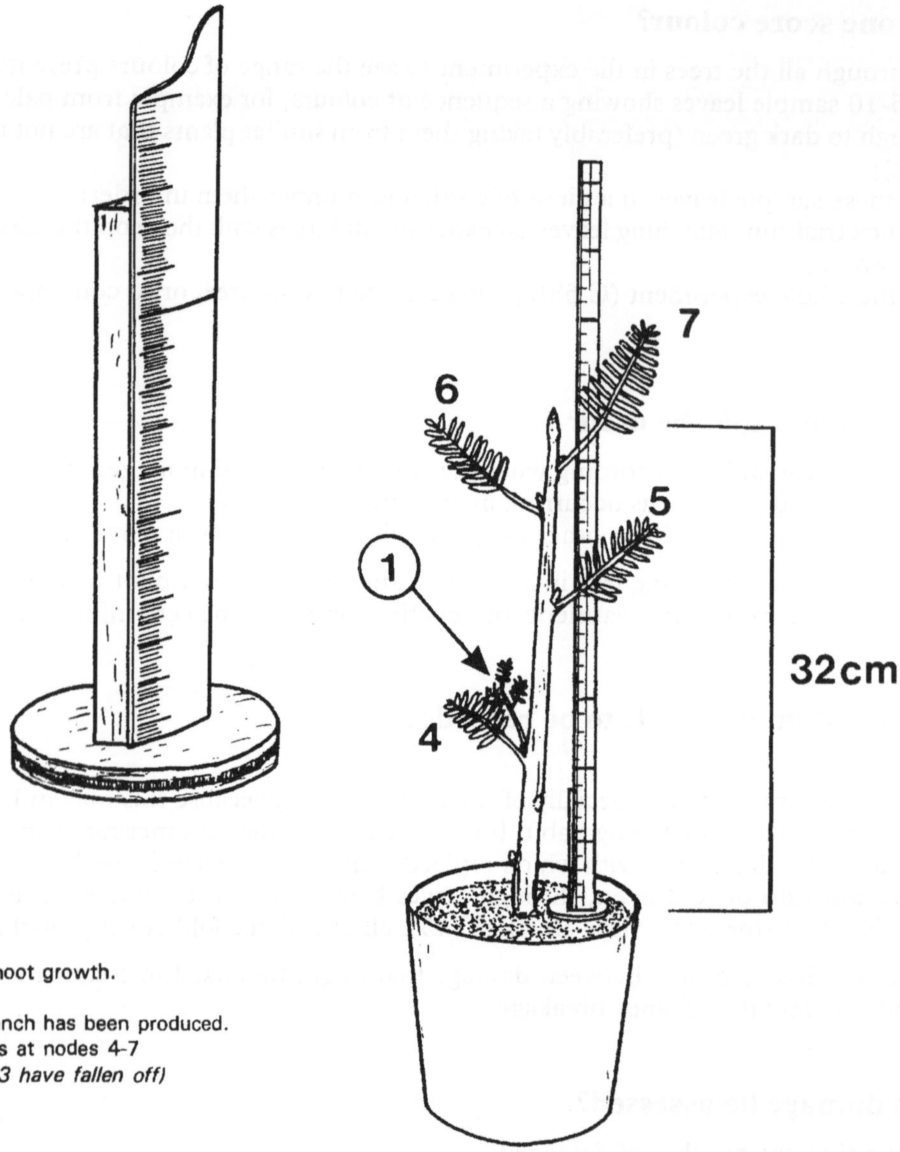

Assessing shoot growth.

①– one branch has been produced.
4-7 – leaves at nodes 4-7
(the bottom 3 have fallen off)

Why use the *gain* in height?

Because the *total* height of the trees includes the different sizes of the trees when the experiment started, and this variability can make it harder for real treatment and/or genetic differences to show up.

Similarly, using the gain in diameter, numbers, length and dry weight can remove a lot of variation, making *significant* results more likely (C 69).

Sometimes the amount of growth relative to the initial size is studied.

How accurate do measurements need to be?

To the nearest centimetre for height and length (or millimetre for detailed measurements);
To the nearest millimetre for diameter and girth (or 0.1 mm for detailed measurements).

What is the easiest way of measuring height?

Usually with a 'fixed' ruler made of suitable wood, plastic or metal. This should be marked in mm or cm, and might be 15, 30, 50 or 100 cm long, whichever is the most convenient. Other alternatives include:

(a) folding or roll-up rulers; ***or***
(b) straight bamboo canes with intervals carefully marked on them with a waterproof marker or paint.

Where should I measure from and to?

(A) **Choosing the base:** measuring from the ground surface is easy, but can be inaccurate because the soil may settle or be washed out, and the ruler may dig in.

For detailed measurements you could:

(1) put in a plastic golf tee or similar object, so that you always measure from the same firm point;

(2) attach a small flat disc to the bottom of the ruler, making sure that it gives a correct measurement from zero; ***or***

(3) measure from a mark made with a waterproof marker or paint at a convenient point on the stem. (*Note: this mark must be on tissue that has stopped elongating* (C 10.))

(B) **Choosing the tip:** it is best to measure to ***where you estimate the growing point of the main stem to be.*** Avoid measuring to the highest point, as this may be part of a leaf or a branch, which will make your data inaccurate and also more variable.

Is it easier to measure diameter or girth?

It is difficult to use a girthing tape on small trees because:

(a) the tape is hard to get straight for reading the girth;

(b) there are often leaf-bases and sometimes branches in the way.

Diameter is usually easier to measure, but getting accurate data depends on:

(1) having calipers of reasonable quality;

(2) not pressing the instrument into soft tissues;

(3) measuring at a standard distance up the stem, while avoiding places where there are irregularities such as the bases of branches; ***and***

(4) marking the stem so that subsequent readings can be made at the same height and on the same sides of the stem.

Is it easy to record the numbers of leaves?

Leaves are surprisingly difficult to keep track of, because:

(a) new leaves may appear in flushes rather than regularly;

(b) some nodes (C 12) may not carry foliage leaves;

(c) older leaves may fall off or be eaten between assessments.

What are the alternatives?

Method A - simply record the number of leaves present on the given date: ***or***

Method B - mark the main stem at a convenient point at the start of the experiment, and count the number of visible leaves and nodes, both towards the tip and towards the base. Then when you make the same assessment later, you can calculate for each tree:

(1) how many new leaves it has produced altogether;

(2) how many have gone;

(3) what is the current number of leaves;

(4) how many new nodes have been formed; ***and***

(5) what is the average length of new internodes.

Method B also allows you to select specific sample leaves (or internodes) of **comparable age** for more detailed study of any treatment or genetic differences.

Why measure the length of leaves?

(a) Because it is easy to do; ***and***

(b) Since the length of the leaf *blade* often turns out to have a close relationship with its area, which is likely to influence the potential amount of photosynthesis (C 12) and the overall growth of the young tree.

If you need to know leaf areas, they can be measured by:

(1) drawing round the leaf on squared paper, counting the number of squares that are at least half included, and multiplying by the size of a square; ***or***

(2) using a *leaf-area meter* with such pieces of paper or with detached leaves.

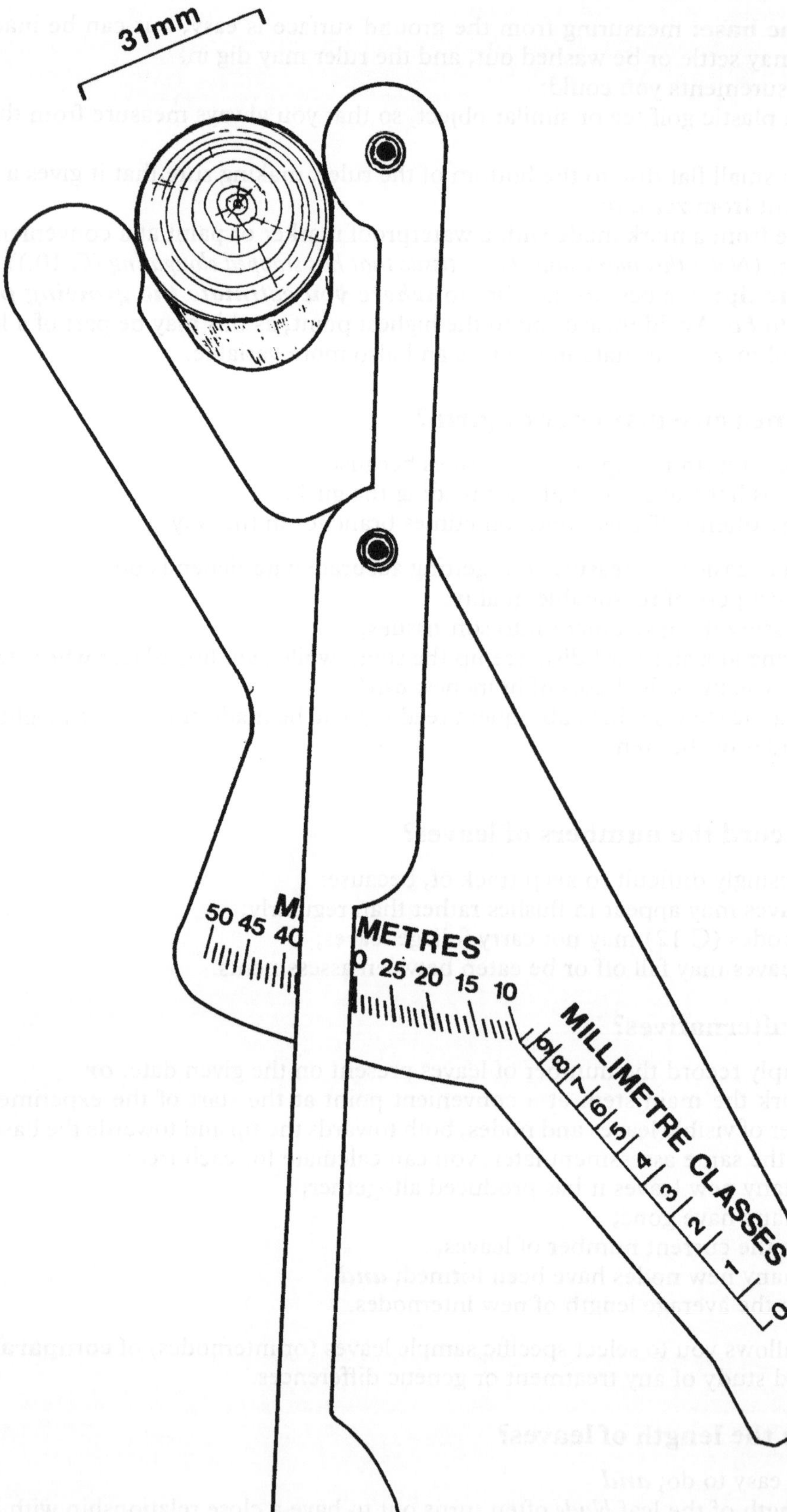

Calipers designed by U.K. Forestry Commission Research Branch.

Would branches need to be measured?

Branch lengths may sometimes be useful, if they can be measured without damage.

How about root growth?

Although this is very important, it is difficult to assess accurately without damaging the young trees, except for:

(a) experiments on rooting of cuttings (A 45 in Manual 1); ***and***

(b) germination tests (Manual 2);

where, with little or no damage at these early stages, you could:

(1) note the first appearance of roots, and how many there are; ***and***

(2) measure the total length of root that has been produced.

Won't destructive sampling disturb the results?

Yes, it can do, particularly if:

(a) the numbers of trees in each treatment are limited;

(b) several other trees were lost during the experiment; ***or***

(c) random sampling happens to remove, for instance, some of the largest trees from one of the treatments.

In addition, as many plants as possible may be needed later on for:

(1) continued measurement;

(2) planting in a field trial; ***or***

(3) using as stockplants (A 6 in Manual 1).

What other non-destructive measurements can be made of roots?

Rough estimates of root development are sometimes made by:

(A) gently tapping out the intact root ball from a tapered pot (C 6), and counting the number of root tips that can seen without disturbing the soil, or scoring the amount of root growth that is visible (C 68);

(B) planting the young trees in a root observation box that has a sloping glass surface through which the progress of individual roots can be followed *(Note that the glass surface needs to be covered with black polythene sheeting except when observations and measurements are being done, in order to exclude light)*; ***or***

(C) growing the young trees in a mist box, with the roots in air that is continually being filled with small drops of water; or in aerated water as in the 'bubble-bath' system (Manual 1).

Supposing I need to know the dry weights?

Unless plenty of young trees are available, you might consider whether it would be adequate to record:

(A) an initial set, using equivalent plants to those in the experiment; ***and***

(B) a final sample, at the end of the experiment.

If some destructive measurements are to be made, you could take the opportunity of washing out the whole root system carefully, photographing it, scoring the amount and distribution of roots, and finding the dry weight at each successive depth in the soil.

What is the procedure for measuring dry weights?

(a) Wash out the roots from the soil carefully, trying not to break the small ones;

(b) Cut up each tree into separate sections (usually stem, leaf and root). Remove surplus water, and weigh them if you will need fresh weights to calculate moisture contents.

(c) Dry batches in an oven at a chosen temperature (generally between 80° and 105°C).

(d) Cool in a *dessicator* to prevent moisture being re-absorbed.

(e) Weigh some of the bulkier samples, dry them again, and weigh again, until a constant weight is recorded. Then dry the remaining batches for a standard period that will have brought them to a constant weight.

Samples of known fresh or dry weights can then be used for any chemical analyses.

Why should I record what happens to the growing conditions?

In some cases this is unnecessary;
In most experiments it can prove useful, because:

(a) a change in weather may influence the rate at which your experimental trees grow;
(b) if some or all of your plants had suffered from wilting, this could be a possible explanation for sudden fluctuations in growth rate or loss of leaves; ***or***
(c) if pests or diseases were noted, these might also have had effects on growth rates.

In detailed studies, measurement of the environment may be an integral part of the research. Sensors and automatic recorders are now available which can provide a continuous record of temperature, light and humidity, for example.

How should I use the measurements of growth that I make?

Even if the experiment has a complex layout, it is usually best to work out the main averages as you go along, and if possible draw some simple graphs. These might suggest for example that:

(1) treatment or genetic differences are starting to have an effect;
(2) measuring needs to be done more often, or less frequently;
(3) observations or scoring could now be relevant, for example if shoot growth seemed to be slowing down, or restarting;
(4) it may soon be time to stop the experiment.

Computers can do many things these days, but leaving all the calculations until afterwards means that you may remain unaware of important aspects until it is too late to respond.

What about analysing the data?

Some advice on statistical analyses is given in sheets C 67-69.

CHECK-LISTS, SOURCES AND RECORDS

Check-lists, sources and records C 60
- *check-list for nursery trees not growing well*

Feature	Possible reasons for poor growth	See sheet number
LEAVES		
twisted	Aphids attacking plant?	C 45
	Leaf damaged earlier by drought or insects?	C 13, 41, 45
pale colour	Soil lacks nutrient(s)?	C 14, 33
	Nutrients unbalanced?	C 14, 33
	Natural feature of new leaves - colour develops later?	C 12
very small	A feature of the particular genetic origin?	C 5
	Young tree has branched a lot?	C 5, 12, 55
	Not enough shade?	C 41
	Previous water stress?	C 13, 25, 41, 46
	Shortage of nutrients?	C 14, 33
prone to wilt	Shade reduced too quickly?	C 41, 47
	Soil too rich and leaves too big?	C 6, 23, 33, 34
	Roots damaged in potting or transplanting?	C 4, 40, 42
	Roots have not had time to grow into new soil?	C 42, 47
	Roots attacked by pest or disease?	C 45
fall off early	Sudden change in environment?	C 4, 25, 40, 46
	Too rapid hardening?	C 47
	Pots too small?	C 6, 63
	Watering problems?	C 43, 52
	Plants attacked by pest or disease?	C 45
have holes	Caterpillars or leaf-miners?	C 45
	Leaf-cutting ants?	C 45
torn	Large, soft leaves damaged by wind?	C 25, 45
	Careless handling of plants?	C 40, 50, 52
	Birds taking insects?	C 45
STEMS		
spindly	Plants too close to each other?	C 42
	Shade too heavy?	C 41
	Pots too small?	C 6
	Poor potting mix?	C 6
	Competition from weeds?	C 44
growth stopped	The species grows in height by periodic flushing?	C 12, 55
	Soil unsuitable?	C 6, 20, 23
	Plants short of nutrients?	C 14, 33
	Watering problems?	C 43
	Environment too shady?	C 41
bent	Genetic characteristic of the species?	C 5
	Temporary feature of young shoots?	C 12
	Edge plant with one-sided foliage?	C 7, 48
	Not enough shelter from wind?	C 25, 46

STEMS (continued)

die-back of tip	Soil waterlogged?	C 6, 20, 23, 67
	Previous severe water stress?	C 13, 41, 55
	Insect attack?	C 45
broken tip	Careless handling?	C 40, 50
	Severe storm?	C 3, 25
	Stem-boring insects?	C 45
	Large animal in nursery?	C 3, 25, 46
forking	Natural feature of the species or clone?	C 5
	Birds attacking buds?	C 45
	Response to previous die-back or breakage?	C 55

ROOTS

pot-bound	Containers too small?	C 6
	Tree too long in same pot?	C 6, 42
few seen	Damaged in potting up?	C 42
	Unsuitable pH of potting mix or bed?	C 6, 23
	Most roots are in the ground beneath the pot?	C 4, 41, 42
many dead	Soil poorly aerated or waterlogged?	C 6, 23
	Root disease?	C 45
	Nematodes or other pest damaging them?	C 45
small clusters along roots	Beneficial nitrogen-fixing nodules?	C 32
	Root aphis or similar pest?	C 45
fine threads near roots	Beneficial mycorrhizal fungus?	C 31
	Harmful fungus?	C 45

WHOLE PLANT

stunted	Seed-lot with inbreeding depression?	C 5
	Shoot development is naturally slower than root growth?	C 11, 12
	Plants need to be repotted into larger containers?	C 6, 42
	Different potting mix or nurery soil required?	C 6
	Shortage of a nutrient?	C 14, 33
	Inoculation for mycorrhizas or nodules needed?	C 30, 31, 32
	Altered shading required?	C 41
trees dying	Unsuitable species or provenance?	C 5
	Poor clone?	C 5
	Unfavourable nursery environment?	C 3, 4
	Insufficient care of young trees?	C 40, 50, 52
	Virus disease?	C 45, 53
	Serious pest?	C 45, 53

GENERAL APPROACH

Unless the batch is specially valuable, break up the soil on a few sample plants and examine the roots, root-collar, stems and leaves in detail, to find out what may be wrong with them.

Suggested action to be taken	Sheet number

For unsuitable genetic origins: C 5

(A) Try several different provenances, local seed sources, or clones;
(B) Avoid collecting seed from single trees or very small groups.

Difficult nursery soils: C 23

Work into the topsoil of seed and transplant beds, *if the nursery soil is:*
- (a) *too acid,* either ground limestone, lime or a fertiliser that increases the pH;
- (b) *alkaline,* add flowers of sulphur or a fertiliser that lowers the pH;
- (c) *too heavy,* work in some sharp sand;
- (d) *too sandy,* work in some silt and extra organic matter;
- (e) *too hard,* try cultivation and extra organic matter, sand or forest topsoil;
- (f) *too wet,* try digging drains and using raised nursery beds;
- (g) *too dry,* try sunken beds and mulching.

Container problems: C 6

(1) Try out pots of different size, shape or type;
(2) Consider using 'root-trainers';
(3) Make up a different potting mix;
(4) Stand the containers on a different surface.

Unsuitable potting mixtures: C 6, 30, 42

(1) *If drainage is poor,* add more sharp sand or grit, check the holes in the containers, and also the standing ground;
(2) *If pots drain too freely,* add more organic matter and finer components, and consider whether the young trees should be potted up more firmly;
(3) *If the pH is unsuitable,* treat the potting mixture as for nursery soils above, but avoiding too much of any nutrient;
(4) *If a crust of algae forms on the surface,* use a mixture less rich in nutrients, and break up the crust with a sharpened stick;
(5) *If a micro-nutrient is lacking,* add some good topsoil or sieved compost or a small amount of a suitable fertiliser;
(6) *If the species has a close association with a micro-organism,* mix in chopped roots and topsoil from under a thriving stand of the same species, nursery soil from a bed where it has grown well, or a special inoculum if this is available.

Problems with watering: C 24, 40, 43, 66

(A) Check water purity and availability, and install a reserve supply;
(B) Check the type of container, potting mix and potting up techniques;
(C) Increase the spacing of plants to allow water to reach each pot;
(D) Consider changing the way the water is put on;
(E) Explain the desired watering method more clearly, and check that it is adopted.

Difficulties over light and shade: C 41, 47, 48

(a) Put the young trees under temporarily heavier shade when their root systems have been disturbed, but avoid keeping them in deep shade too long;
(b) Reduce shading gradually, rather than suddenly, allowing for changing seasons;
(c) Consider planting suitable shade trees.

Shelter problems: C 25, 46, 48

(A) Plant hedges to check the force of the wind and reduce water loss by the young trees;
(B) Raise small seedlings and delicate species under a roof to protect them from heavy raindrops;
(C) Consider building a shadehouse or simple greenhouse.

Dealing with insect pests: **C 45, 61-C**

(1) Look out for early signs of a pest attack, including a check on the roots of the trees;
(2) Take off or squash the insects, and check again every day or two;
(3) Remove any nearby plants or weeds that may be acting as a centre for spread;
(4) Spray affected trees with water containing a little detergent, or if necessary with an insecticide.

Other pest problems: **C 45, 61-C**

(a) Try to find out what kind of pest is present, and how it can be controlled;
(b) Follow (1-4) above, where appropriate;
(c) Avoid spreading the resting stages of the pest to other plants, for instance through topsoil, compost or mulch.

Difficulties due to disease: **C 45**

(A) Try not to have very damp, cool and shady environments, except for poly-propagators;
(B) Remove weeds and rotting leaves regularly;
(C) Look out for signs of problems, such as moulds on young leaves, cankers on stems or roots that are dying;
(D) Increase ventilation and consider decreasing shade and watering;
(E) Search for a book or someone who can identify what kind of micro-organism is responsible;
(F) If necessary, spray with a fungicide, or add one to the potting mix.

Careless handling of plants: **C 40, 50, 52**

(1) Explain the reasons for handling young trees carefully;
(2) Correct anyone who is treating them roughly;
(3) If work is paid by the amount done, consider changing to a system with a bonus for the number of good plants raised.

Other problems:

(a) For poorly rooted cuttings, see sheets A 2, A 50 and A 61 in Manual 1;
(b) For poor seed germination, see Manual 2.

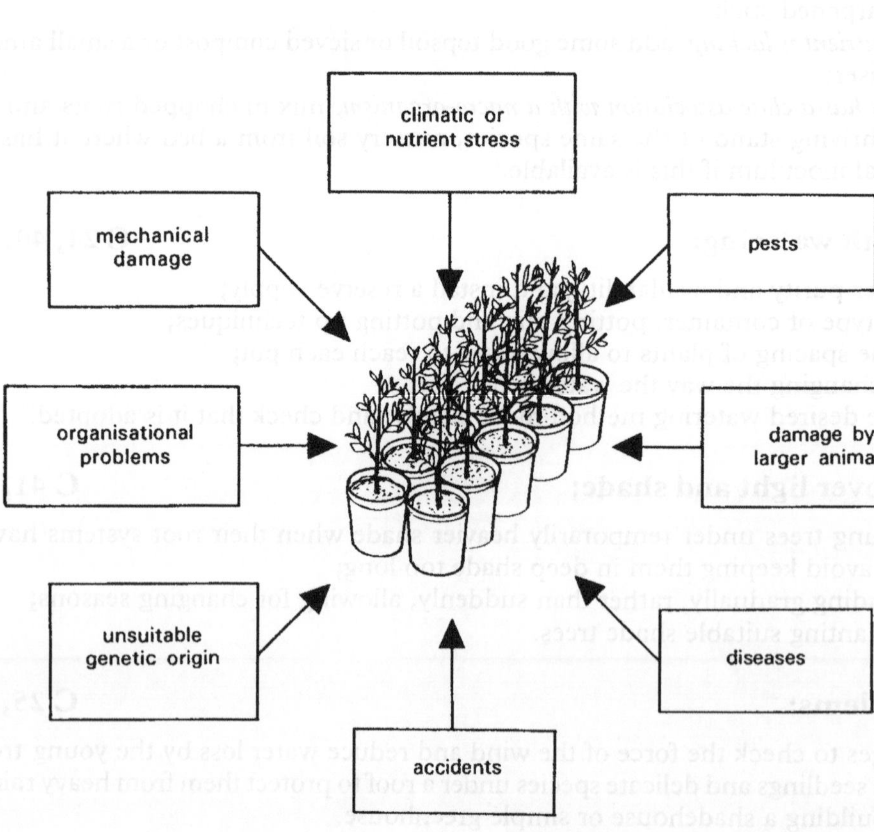

Check-lists, sources and records C 61
- *some information about nurseries*

(A) Tree nursery manuals:

Anonymous (1986). *Manuel de viveros y plantaciones escolares.* Ministerio de Educacion/ Ministerio de Agricultura, Lima, Peru.

Anonymous (1987). *El establecimiento y manejo de un pequeno vivero forestal.* Banco Nacional de Costa Rica/El Cuerpo de Paz de los Estados Unidos.

Anonymous (1996). *Field guide on improved nursery technology,* Andhra Pradesh Forest Department, Hyderabad, India.

Arroyo, G.S., Bolanos, S.R. and Murillo, L.F. (1989). *Manuel metodologico viveros forestales.* Proyecto Cooperativo de Educacion Forestal, San Jose, Costa Rica.

Carter, E.J. (1987). *From seed to trial establishment. A handbook giving practical guidelines in nursery practice and the establishment of simple species and/or provenance trials.* DFR User Series, **2**, Division of Forest Research, Commonwealth Scientific and Industrial Research Organization, P.O. Box 4008, Yarralumla, ACT 2600, Australia.

Davey, C.B. (1984). *Pine nursery establishment and operation in the American tropics.* Central America and Mexico Resources Cooperative, North Carolina State University, Bulletin **1**.

Dirr, M.A. and Heuser, C.W., Jr. (1987). *The reference manual of woody plant propagation.* Varsity Press, Inc., P.O. Box 6301, Athens, Georgia, USA.

Dupriez, H. and de Leener, P. (1983). *Jardins et vergers d'Afrique.* Terres et Vie, Nivelles, Belgium.

Dupriez, H. and de Leener, P. (1992). *African gardens and orchards.* Macmillan, London, England.

Duryea, M.L. and Landis, T.D. (Eds.) (1984). *Forest Nursery Manual: production of bareroot seedlings.* Martinus Nijhoff/W. Junk, The Hague, Netherlands.

Fuller, B.R. (1983). *Planting and managing kraal woodlots.* Forestry Commission, Harare, Zimbabwe.

Gerber, H.L. and Herbst, F.A. (1982). *Tree nursery handbook.* Department of Environmental Affairs, Pretoria, S. Africa.

Goeltenboth, F. (Ed.) *et al.* (1990). *Subsistence agriculture improvement: manual for the humid tropics.* Wau Ecology Handbook, **10**, 2nd edn, section 4, Wau Ecology Institute, Wau, Papua New Guinea/Tropical Agroecology, **4**, Margraf Scientific Publishers, Weikersheim, Germany.

Goor, A.Y. and Barney, C.W. (1968). *Forest tree planting in arid zones,* Chapter 3: *Forest-tree nurseries,* Ronald Press, New York, USA.

GTZ (1976). *Manual of reforestation and erosion control for the Philippines.* German Agency for Technical Co-operation, P.O. Box 5180, Eschborn, Germany.

ILO (1989). *Tree nurseries: an illustrated technical guide and training manual.* International Labour Organization, Special Public Works Programmes, Booklet **6**.

Liegel, L.H. and Venator, C.R. (1987). *A technical guide for forest nursery management in the Caribbean and Latin America.* Forest Service, U.S. Dept. of Agric., Gen. Tech. Report **SO-67.**

Mung'ala, P.M., Kuyper, J.B.H. and Kimwe, S. (1988). *On-farm tree nurseries.* Kenya Woodfuel Development Programme, Nairobi, Kenya.

Napier, I. (1985). *Técnicas de viveros forestales con referencia especial a Centroamérica.* Publicación Miscelanea **5**, Escuela Nacional Ciencias Forestales, Siguatepeque, Honduras, Central America.

Napier, I and Robbins, M. (1989). *Forest seed and nursery practice in Nepal.* Nepal/UK Forestry Research Project, Dept of Forestry and Plant Research, Babar Mahal, P.O. Box 3339, Kathmandu, Nepal.

Napier, I. and Willan, R.G. (1983). *Nursery techniques for tropical pines.* Danida Forest Seed Centre, Humlebæk, Denmark, Technical Note **3.**

Nieuwenhuis, J. and Hussain, R. (1990) *Nursery techniques training manual.* Malakand Social Forestry Project, NWFP Forest Department, Pakistan.

Pancel, L. (1993). Nursery management. Pp 464-514 in *Tropical forestry handbook,* ed. Pancel, L., Volume **1,** Springer, Heidelberg, Germany.

Thunberg, J. (1984). *Village nurseries for forest trees: how to set them up and how to run them.* Swedish International Development Authority (SIDA), Stockholm.

Tinus, R.W. and McDonald, S.E. (1979). *How to grow tree seedlings in containers in greenhouses.* Forest Service, U.S. Dept. of Agric., General Technical Report **RM-60.**

Weber, F.R. (1986). *Reforestation in arid lands.* 2nd edition, VITA, 1815 North Lynn St, Suite 200, Arlington, Virginia, USA.

Zaïre/Canada (1988). *La production de semis en pepinière* Projet Pilote d'appui aux Reboisements Communitaires, Publ. **88/2** (in French and Kikongo), Kinshasa, Zaïre.

Zaïre/Canada (1990). *Disposition, entretien et protection des plantations.* Projet Pilote d'appui aux Reboisements Communitaires, Publ. **90/1** (in French and Kikongo), Kinshasa, Zaïre.

(B) Nursery tools, containers, materials and equipment:

Appleton, B.L. (1993). Nursery production alternatives for reduction or elimination of circling tree roots. *Journal of Arboriculture,* **19,** 383-388.

Dalzell, H.W., Biddlestone, A.J., Gray, K.R. and Thurairajan, K. (1987). *Soil management: compost production and use in tropical and subtropical environments.* FAO Soils Bulletin **56,** Rome, Italy.

Dubois, J. and Davio, C. (1960). Deux techniques particulières en matière de pepinière forestière tropicale. *Bull. Inform. Inst. nat. Etude agron. Congo,* **9,** 313-330.

Felker, P., Wiesman, C. and Smith, D. (1988). Comparison of seedling containers on growth and survival of *Prosopis alba* and *Leucaena leucocephala* in semi-arid conditions. *Forest Ecology and Management,* **24,** 177-182.

Josiah, S.J. (1988). *Containerised seedling production for reforestation in Haiti.* Seminario de Fincas Energeticas, Establecimiento y Actualizacion de Costos, Santo Domingo, Dominican Republic.

Josiah, S.J. and Jones, N. (1992). *Root trainers in seedling production systems for tropical forestry and agroforestry.* World Bank, Washington D.C., USA.

Kijkar, S. (1991). *Coconut husk as a potting medium.* ASEAN-Canada Forest Tree Seed Centre Project Handbook, Muak-Lek, Saraburi 18180, Thailand.

Landis, T.D., Tinus, R.W., McDonald, S.E. and Barnett, J.P. (1990). *The container tree nursery manual. Vol. II: containers and growing media.* Forest Service, U.S. Department of Agriculture Handbook **674.**

Maurice, J. (1985). Integrated techniques for propagating and conveying compact grafted trees under Third World conditions. *Comb. Proc. Internat. Plant Propagators' Society,* **34,** 177-185.

Miller, J.H. and Jones, N. (1995). Organic and compost based growing media for tree seedling nurseries. *World Bank Technical Paper,* **264,** Washington, D.C., USA.

Muthana, K.D and Harsh, L.N. (1986). Use of super absorbent polymers in arid zone forestry. *J. Tropical Forestry,* **2,** 164-167.

Nwonwu, F.O.C. (1987). An assessment of the suitability of a soil amendment polymer for tree crop growing. *Pakistan Journal of Forestry,* **37,** 191-196.

Ole-Meilude, R.E.L. and Pamba, G.K. (1988). Time study on different techniques for nursery pot filling. *Silva Fennica,* **22,** 171-175.

Reid, R.K. (1989). *Seedling growth and development in different container types and soil mixes.* Haiti Agroforestry Research Project, South-East Consortium for International Development/Auburn University, USA.

Rojas, F. and Rodriguez, F. (1995). Root trainers: an option for the production of seedlings in forest nurseries (*in Spanish*). pp 16-19 in *Planes simplificados de manejo forestal,* eds. Rojas, F. *et al.*, CATIE Revista Forestal Centroamericana **3.**

Sharma, R.D. (1987). Some observations on coiling of roots in nursery raised plants. *Journal of Tropical Forestry,* **3,** 207-212.

Venator, C.R. and Munoz, J.E. (1974). Containerised tree production in the tropics. Pp 334-335 in *Proc. N. American Containerized Forest Tree Seedling Symp.,* Denver, Colorado, USA.

Wilson, P.J. (1986). Containers for tree nurseries in developing countries. *Commonwealth Forestry Review,* **65,** 233-240.

(C) Nursery pests and diseases:

Anonymous (1996). *Using pesticides: a complete guide to safe, effective spraying.* British Crop Protection Council, Bracknell, England.

Brunck, F. (1994). Les ravageurs et maladies de l'Acacia nilotica. *Bois et Forêts des Tropiques,* **239,** 59-65.

Chaplin, G.E. (1987). Insect pests and fungal diseases of trees in the Solomon Islands: recent identifications. *Forest Research Note* **29-1/87,** Ministry of Natural Resources, Solomon Islands.

Chilima, C.Z. (1991). Termite control in young Eucalyptus plantations in Malawi using controlled release insecticides. *Commonwealth Forestry Review,* **70,** 237-247.

NAS (1992). *Neem: a tree for solving global problems.* BOSTID publ. **71,** National Academy of Sciences, Washington, DC, USA.

Gibson, I.A.S. (1975). *Diseases of forest trees widely planted as exotics in the tropicsand Southern Hemisphere.* Part I: *Important members of the Myrtaceae, Leguminosae, Verbenaceae and Meliaceae.* Commonwealth Mycological Institute, Kew, England.

Gibson, I.A.S. (1979). *Diseases of forest trees widely planted as exotics in the tropics and Southern Hemisphere.* Part II: *The genus Pinus.* Commonwealth Mycological Institute, Kew, England.

Schmutterer, H. and Ascher, K.R.S. (Eds) (1987). *Natural pesticides from the neem tree (Azadirachta indica A. Juss.) and other tropical plants.* Proc. 3rd Internat. Neem Conf., Nairobi, Kenya, July 1986, GTZ, Eschborn, Germany.

Sushilkumar, Sen-Sarma, P.K. and Jha, L.K. (1994). Insect pests management for increased forest productivity in India. *Annals of Forestry,* **2,** 190-206.

Wardell, A. and Wardell, D.A. (1990). The African termite: peaceful coexistence or total war? *Agroforestry Today,* **2,** 4-6.

(D) Other information about tree nurseries and propagation:

Ahujia, M.R. (Ed.) (1994). *Micropropagation of woody plants.* Kluwer Academic Publishers, The Netherlands.

Agritex (1990). *Agroforestry extension training manual.* Agricultural Technical and Extension Services, Causeway, Zimbabwe.

Baggio, A. and Heuveldop, J. (1984). Initial performance of *Calliandra calothyrsus* Meissn. in live fences for the production of biomass. *Agroforestry Systems,* **2,** 19-29.

Benge, M.D. (1987). *More tree planting through school nurseries.* USAID Report, Washington, D.C., USA.

Campbell, M.W. (1983). *Plant propagation for reforestation in Nepal.* Nepal-Australia Forestry Project, Technical Note **1/83,** Australian National University, Canberra.

Desmond, D.F. (1989). *Forest tree nurseries in Agricultural High Schools: analysis of Ecuadorean experiences.* Social Forestry Network Paper **9e,** Overseas Development Institute, London, England.

Dick, J.M. and Hamzah, A. (1994). Vegetative propagation of tree species indigenous to Malaysia. *Commonwealth Forestry Review,* **73,** 164-172.

Egli, A. and Kalinganire, A. (1988). *Les arbres et arbustes agroforestiers au Rwanda.* Institut des Sciences Agronomiques, Butare, Rwanda.

Fatimson, T. (1989). *Drylands agroforestry, homestead trees and the nurseries to support them.* Social Forestry Network Paper **9b,** Overseas Development Institute, London, England.

Fearnside, A. and Drew, I.K. (1977). *Fence construction techniques for forestry in Nepal.* Nepal-Australia Forestry Project, Technical Note **1/77,** Kathmandu, Nepal.

Foster, G.S. and Diner, A.M. (Eds) (1994). Applications of vegetative propagation in forestry. Proc. Symposium on forest genetics, in Alabama, 1992, *Southern Forest Experiment Station, General Technical Report,* **SO-108,** U.S. Department of Agriculture.

Hartmann, H.T. and Kester, D.E. (1983). *Plant Propagation: principles and practices,* 4th edn, Prentice Hall, New Jersey, USA.

Jagawat, H. (1989). Observations on centralized and decentralized nurseries in an NGO project in Gujerat, North India. In *Nurseries in Gujerat, North India: two views.* Social Forestry Network Paper **9d,** Overseas Development Institute, London, England.

Khan, S.A. (1985). Nursery Practices. Pp 227-256 in *Increasing productivity of multipurpose trees,* eds Burley, J. and Stewart, J.L., IUFRO, Vienna, Austria.

Mbonye, A. (1988). *How to stop the desert: an NGO action guide on desertification control.* African NGOs Environment Network, P.O. Box 53844, Nairobi, Kenya.

Monteuuis, O. and others (1995). Propagation clonale de tecks matures par bouturage horticole. *Bois et Forêts des Tropiques,* **243,** 25-39.

Nelson, W.R. (1989). A review of the role of root pruning for containerized forest seedlings. *South Africa Forestry Journal* **151,** 90-92.

Piggott, C.J. (1990). *Growing oil palms: an illustrated guide.* Incorporated Society of Planters, Kuala Lumpur, Malaysia.

Poupard, C., Chauvière, M. and Monteuuis, O. (1994). Acacia mangium cuttings: effects of age, within-shoot position and auxin treatment. *Silvae Genetica,* **43,** 226-231.

Robinson, P. and Thompson, I. (1989). *Fodder trees, nurseries and their central role in the hill-farming systems of Nepal.* Social Forestry Network Paper **9a,** Overseas Development Institute, London, England.

Rojas, R.F. (Ed.) (1987). *Primer taller nacional sobre semillas y viveros forestales.* Memoria. Taller realizado en San Jose del 25 al 29 de Noviembre de 1985. CATIE, Turrialba, Cartago, Costa Rica.

Shanks, E. and Carter, J. (1994). *The organisation of small-scale tree nurseries.* Rural Development Study Guide **1,** Overseas Development Institute, London, England.

Verma, D.P.S. (1989). People's decentralised nurseries: field-level experience of the Gujerat Forest Department, India; in *Nurseries in Gujerat, North India:* two views. Social Forestry Network Paper **9d,** Overseas Development Institute, London, England.

von Maydell, H.J. (1983). *Arbres et arbustes du Sahel.* GTZ, Eschborn, Germany.

Westwood, S. (1986). *Results of experiments conducted at Hetauda Forest Research Nursery,* 1984-85. Publ. **45** of Forest Survey and Research Office, Nepal.

White, K.J. (1988). *Forest nursery strategies and practices in the Bhabar Terai of Central Nepal.* Sagarnath Forest Development Project, Manual **3,** 2nd edn, Ministry of Forests, Kathmandu, Nepal.

Notes:

(1) See sheet A 62 in Manual 1 for more information on vegetative propagation, and sheet A 63 for some sources of chemicals and materials;

(2) More information on seed collection, storage, dormancy and germination will be given in Manual 2.

Check-lists, sources and records C 62
- *some information on tree growth*

(A) Effects of treatments on tropical tree growth:

Abod, S.A. and Cheong, K.L. (1994). Effects of a growth retardant and shoot pruning on the growth of Acacia mangium seedlings. *J. Tropical Forest Science,* **6,** 239-248.

Awang, K. and de Chavez, C.G. (1993). Effect of root-wrenching and controlled watering on growth, drought resistance and quality of bare-rooted seedlings of Acacia mangium. *J. Tropical Forest Science,* **5,** 309-321.

Baggio, A. and Heuveldop, J. (1982). *Initial performance of Calliandra calothyrsus Meissm. in live fences for the production of biomass.* CATIE, Turrialba, Costa Rica.

Bowman, D.M.J.S. and Panton, W.J. (1993). Factors that control monsoon-rainforest seedling establishment and growth in North Australian Eucalyptus savanna. *Journal of Ecology,* **81,** 297-304.

Chalot, M., Battut, P.M., Botton, B., Le Tacon, F. and Garbaye, J. (1988). Recent advances in physiological and practical aspects of ectomycorrhizal effects on tree development. *Acta écologica écol. Applic.* **9,** 333-351.

Egli, A. and Kalinganire, A. (1988). *Les arbres et arbustes agroforestiers au Rwanda.* Institut des Sciences Agronomiques, Butare, Rwanda.

Longman, K.A. (1978). Control of shoot extension and dormancy - external and internal factors. Pp 465-495 in *Tropical Trees as Living Systems,* eds Tomlinson, P.B. and Zimmermann, M.H., Cambridge University Press, Cambridge, England.

Longman, K.A. & Jenĭk, J. (1987). *Tropical Forest and its Environment,* 2nd Edn, Longman Scientific & Technical, Harlow, England/John Wiley, New York, USA.

Nasi, R. and Monteuuis, O. (1993). Un nouveau programme de recherches au Sabah. *Bois et Forêts des Tropiques,* 235, 25-34.

Rauch, F.D. (Ed.) (1989). *Hawaii nursery research.* Hawaii Inst. of Tropical Agric. and Human Resources, Research Extension Series, **103.**

Singh, A.K., Meena, S.L., Williams, A.J., Banerjee, S.K. and Gupta, B.N. (1994). Effect of mulches and watering interval on growth of *Albizia procera in* skeletal soil. *Environment and Ecology,* **12,** 93-98.

Wright, S.J. (1996). Phenological responses to seasonality in tropical forest plants. Pp 440-460 in *Tropical forest plant ecophysiology,* eds Mulkey, S.S., Chazdon, R.L. and Smith, A.P., Chapman and Hall, New York, USA.

(B) Soils and roots:

Cottenie, A. (1980). Soils and plant testing as a basis of fertizer recommendations. *Soils Bulletin,* **38** (2), p. 19, FAO, Rome Italy.

FAO (1978). Organic materials and soil productivity, *Soils Bulletin,* **35,** FAO, Rome, Italy.

IITA (1979). *Selected methods for soil and plant analysis.* Manual Series **1,** International Institute for Tropical Agriculture, Ibadan, Nigeria.

Jeník, J. (1978). Roots and root systems in trees: morphologic and ecologic aspects. Pp 332-349 in *Tropical Trees as Living Systems,* eds Tomlinson, P.B. and Zimmermann, M.H., Cambridge University Press, Cambridge, England.

Lal, R. and Greenland, D.J. (Eds) (1979). *Soil physical properties and crop production in the tropics.* Wiley, Chichester, England.

Timberlake, J.R. and Calvert, G.M. (1993). *Preliminary root atlas for Zimbabwe and Zambia.* Drawings by J.A. Morris. Zimbabwe Forestry Commission, Bull. of Forest Research **10.**

Young, A. (1986). Effects of trees on soils. Pp 28-41 in *Amelioration of soil by trees.* Commonwealth Science Council, London, England.

(C) Mycorrhizas:

Adholeya, A. and Khanna, S. (1989). VA mycorrhizae: an important component of tree farming. *J. Tropical Forestry,* 5,17-29.

Alexander, I.J. (1989). Mycorrhizas in tropical forests. Pp 169-188 in *Mineral nutrients in tropical forest and savanna ecosystems,* ed. Proctor, J., Blackwell, Oxford, England.

Alexander, I, Ahmad, N. and See, L.S. (1992). The role of mycorrhizas in the regeneration of some Malaysian forest trees. *Phil. Trans. Royal Society of London B,* **335,** 379-388.

Bulakali, B., Khasa, P.D. and Luyindula, N. (1992). Effets de la double symbiose Rhizobium-Glomus spp. sur la croissance de Leucaena leucocephala (Lam.) de Wit. en pepinière, et dix mois après transplantation au Zaïre. *Tropicultura,* **10,** 132-136.

Chipompha, N.W.S. (1989). Low-cost Pisolithus mycorrhizal inoculum for small-scale pine nurseries. Pp 319-322 in *Trees for development in Sub-Saharan Africa,* Proc. Regional Seminar at ICRAF, Nairobi, Kenya, published by Internat. Foundation for Science, Stockholm, Sweden.

Feldmann, F. and others (1995). *Recultivation of degraded, fallow lying areas in central Amazonia with equilibrated polycultures: response of useful plants to inoculation with VA-mycorrhizal fungi.* Proc. Symp. on Tropical Trees, Hamburg, Germany, September 1993, published in *Angewandte-Botanik,* **69,** 113-118.

Janos, D.P. (1983). Tropical mycorrhizas, nutrient cycles and plant growth. Pp 327-345 in *Tropical rain forest: ecology and management.* Blackwell, Oxford, England.

Janos, D.P. (1985). Mycorrhizal fungi: agents or symptoms of tropical community composition? Pp 98-103 in *Proc. 6th N. American Conf. on Mycorrhizae,* ed. Molina R., Oregon State University, Corvallis, USA.

Lapeyrie, F. and Högberg, P. (1994). Harnessing symbiotic associations: ectomycorrhizas. Pp 158-164 in *Tropical trees: the potential for domestication and the rebuilding of forest resources,* eds Leakey, R.R.B. and Newton, A.C., Institute of Terrestrial Ecology, Symposium **29/** Edinburgh Centre for Tropical Forests, Symp. **1,** Edinburgh, Scotland.

Martinez-Amores, E., Valdes, M. and Quintos, M. (1990). Seedling growth and ectomycorrhizal colonization of Pinus patula and P. radiata inoculated with spores of Helvella lacunosa, Russula brevipes or Lycoperdon perlatum. *New Forests,* **4,** 237-245.

Mason, P.A. and Wilson, J. (1994). Harnessing symbiotic associations: vesicular-arbuscular mycorrhizas. Pp 165-175 in *Tropical trees: the potential for domestication and the rebuilding of forest resources,* eds Leakey, R.R.B. and Newton, A.C., Institute of Terrestrial Ecology, Symp. **29/**Edinburgh Centre for Tropical Forests, Symp. **1,** Edinburgh, Scotland.

Michelson, A. (1993). Growth improvement of Ethiopian acacias by addition of vesicular-arbuscular mycorrhizal fungi or roots of native plants to non-sterile nursery soil. *Forest Ecology and Management,* **59,** 193-206.

Sanders, F.E., Mosse, B. and Tinker, R.B. (Eds) (1975). *Endomycorrhizas.* Proc. Symp. at Leeds, England, July 1974, Academic Press, London, England.

Verma, R.K., Jamaluddin and Gupta, B.N. (1994). Effect of inoculation of VAM fungi and Rhizobium on growth and biomass production in Acacia nilotica in nursery. *Indian Forester,* **120,** 1089-1094.

(D) Nitrogen-fixing tree and shrub species:

Anonymous (undated). *Vegetative propagation of Casuarina equisetifolia.* Andhra Pradesh Forest Department, Hyderabad, India.

Allen, O.N. and Allen, E.K. (1981). *The Leguminosae - a source book of characteristics, uses and nodulation.* University of Wisconsin Press, Madison, Wisconsin, USA.

Aronson, J., Ovalle, C. and Avendaño, J. (1992). Early growth rate and nitrogen fixation potential in forty-four legume species grown in an acid and a neutral soil from central Chile. *Forest Ecology and Management,* **47,** 225-244.

Blom, P.S. (1981). Leucaena - a promising versatile leguminous tree for the tropics. *Internat. Tree Crops J.,* **1,** 221-234.

Bulakali, B., Khasa, P.D. and Luyindula, N. (1992). Effets de la double symbiose Rhizobium-Glomus spp. sur la croissance de Leucaena leucocephala (Lam.) de Wit. en pepinière, et dix mois après transplantation au Zaïre. *Tropicultura,* **10,** 132-136.

Deans, J.D., Ali, O.M., Lindley, D.K. and Nour, H.D.A. (1993). Rhizobial inoculation of Acacia tree species in Sudan: soil inoculum potential and the effects of peat. *J. Tropical Forest Science,* **6,** 56-64.

Dommergues, Y.R. (1987). The role of biological nitrogen fixation in agroforestry. Pp 245-271 in *Agroforestry: a decade of development,* eds Steppler, H.A. and Nair, P.K.R., ICRAF, Nairobi, Kenya.

Faria, S.M. de, Lewis, G.P., Sprent, J.I. and Sutherland, J.M. (1989). Occurrence of nodulation in the Leguminosae. *New Phytologist,* **111,** 607-619.

Franco, A. A. (1984). Fixacao de nitrogenio em arvores e fertilidade do solo. In *Nutrition of Leguminous Trees,* eds Franco, A. A. and others, *Pesquisa Agropecuaria Brasileira* **19** (special issue).

Froidevaux, L. (1985). *Nodulation de quelques arbres fixateurs d'azote au Rwanda.* Institut des Sciences Agronomiques, Note Technique **2,** Rubona, Rwanda.

Giller, K. and Wilson, K. (1993). *Nitrogen fixation in tropical cropping systems.* Intercept, Andover, England.

Gordon, J.C. and Wheeler, C.T. (Eds) (1983). *Biological nitrogen fixation in forest ecosystems: foundations and applications,* Martinus Nijhoff, Dordrecht, The Netherlands.

Lindsay Falvey, J. (1982). Gliricidia maculata - a review. *Internat. Tree Crops J.,* **2,** 1-14.

Nair, P.K.R, Fernandes, E.C.M. and Wambugu, P.N. (1984). Multipurpose leguminous trees and shrubs for agroforestry. *Agroforestry Systems,* **2,** 145-163.

NAS (1983a). *Mangium and other fast-growing acacias for the humid tropics.* BOSTID publ. **41,** National Academy of Sciences, Washington, D.C., USA.

NAS (1983b) *Calliandra: a versatile small tree for the humid tropics.* BOSTID publ. **42,** National Academy of Sciences, Washington, D.C., USA.

NAS (1983c). *Casuarinas: nitrogen-fixing trees for adverse sites.* BOSTID publ. **43,** National Academy of Sciences, Washington, D.C., USA.

NAS (1984). *Leucaena: Promising forage and tree crop in developing countries.* BOSTID publ. **52,** National Academy of Sciences, Washington, D.C., USA.

Pinyopusarerk, K. and House, A.P.N. (1994). *Casuarina: an annotated bibligraphy of C. equisetifolia, C. junghuhniana and C. oligodon,* Division of Forestry, CSIRO, Canberra, Australia.

Reddell, P. (1990) Increasing productivity in plantings of Casuarina by inoculation with Frankia. Pp 133-140 in *Advances in Casuarina Research and Utilization.,* eds El-Lakany, M.H., Turnbull, J.W. and Brewbaker, J.L., American University in Cairo, Egypt.

Reddell, P., Rosbrook, P.A. and Ryan, P.A. (1989). Managing nitrogen fixation in Casuarina species to increase productivity. Pp 209-214 in *Trees for the tropics: growing Australian multipurpose trees and shrubs in developing countries,* ed. Boland, D.J., Australian Centre for International Agricultural Research, Monograph **10,** Canberra, Australia.

Sprent, J.I. (1994). Harnessing symbiotic associations: the potentials for nitrogen-fixing trees. Pp 176-182 in *Tropical trees: the potential for domestication and the rebuilding of forest resources,* eds Leakey, R.R.B. and Newton, A.C., Institute of Terrestrial Ecology, Symp. **29/** Edinburgh Centre for Tropical Forests, Symp. **1,** Edinburgh, Scotland.

Verma, R.K., Jamaluddin and Gupta, B.N. (1994). Effect of inoculation of VAM fungi and Rhizobium on growth and biomass production in Acacia nilotica in nursery. *Indian Forester,* **120,** 1089-1094.

Westley, S.B. and Powell, M.H. (1993). Erythrina in the new and old worlds. *Nitrogen Fixing Tree Association, Reports* (special issue).

(E) Genetic improvement and genetic conservation of trees:

Butterfield, R.P. (1995). Promoting biodiversity: advances in evaluating native species for reforestation. *Forest Ecology and Management,* **75,** 111-123.

de Zoysa, N.D. and Ashton, P.M.S. (1991). Germination and survival of Shorea trapezifolia: effects of dewinging, seed maturity, and different light and soil microenvironments. *J. Tropical Forest Science,* **4,** 52-63.

Drysdale, R.M., John, S.E.T. and Yapa, A.C. (Eds) (1994). Proc. internat. Symp. *Genetic conservation and production of tropical forest tree seed,* Chiang Mai, 1993, ASEAN Tree Seed Centre, Muak Lek, Saraburi 18180, Thailand.

FAO/IPGRI (1994). *Genebank standards.* FAO and Internat. Plant Genetic Resources Institute, Rome, Italy *(also published in French and Spanish).*

Guarino, L., Rao, V.R. and Reid, R. (1994). *Collecting plant genetic diversity: technical guidelines.* Commonwealth Agricultural Bureaux International, Wallingford, Oxford, England.

Hattemer, H.H. and Melchior, G.H. (1993). Genetics and its application to tropical forestry. In *Tropical Forestry Handbook,* ed. Pancel, L., Springer Verlag, Berlin, Germany.

Hughes, C.E. (1994). Risks of species introductions in tropical forestry. *Commonwealth Forestry Review,* **73,** 243-252.

ITTO (1994). Guidelines on the conservation of biological diversity in tropical production forests. *ITTO Policy Development Series,* **5,** International Tropical Timber Organization, Yokohama, Japan.

Jiminez, V. and Picado, W. (1987). Algunas experiencias co Acacia mangium en Costa Rica. *Silvoenergia,* **22.**

Jon-Llap, R., Camacho-Hernandez, Y., Viquiez, E. and Sanchez, G. (1990). Comportiamento juvenil de procedencias y familias de Gliricidia sepium de la region de origen, *Chasqui-Turrialba,* **22,** 7-13.

Leakey, R.R.B. and Newton, A.C. (Eds) (1994). Domestication of tropical trees for timber and non-timber products. *MAB Digest,* **15,** UNESCO, Paris, France.

Park, Y.G. and Sakamoto, S. (Eds) (1994). Proc. internat. Seminar *Genetic conservation and utilization of biodiversity,* Inst. Agric. Science and Technology, Kyungpook National University and Genetic Resources Center, Taegu 702-701, Korea.

Salazar, R. and Vasquez, M.S. (1988). Variacion genetica en ocho procedencias de Erythrina poeppigiana en Costa Rica. *Turrialba,* **38,** 71-81.

Salazar, R. (1989). Genetic variation of 16 provenances of Acacia mangium at nursery level in Turrialba, Costa Rica. *Commonwealth Forestry Review,* **68,** 263-272.

Sollart, K. (1986). *The Javanese mixed home garden as a plant genetic resource.* Fakultatas Kehetanan Universitas Gadjah Mada, Jakarta, Indonesia.

Wilcox, B.A. (1995). Tropical forest resources and biodiversity: the risk of forest loss and degradation. *Unasylva,* **46,** 43-49.

Willan, R.L. (1984). *Provenance seed stands and provenance conservation stands.* Technical Note **14,** Danida Forest Seed Centre, Humlebæk, Denmark.

(F) Layouts, assessment and statistical analysis of experiments:

Fisher, R.A. and Yates, F. (1974). *Statistical tables for biological, agricultural and medical research,* 6th edn, Longman Group, Harlow, England.

Heath, D. (1995). *An introduction to experimental design and statistics for biology.* University College of London Press.

Moroney, M.J. (1956). *Facts from figures,* Penguin Books, London, England.

Parker, R.E. (1973). *Introductory statistics for biology.* Studies in Biology, **43,** Edward Arnold, London.

Williams, E.R. and Matheson, A.C. (1994). *Experimental design and analysis for use in tree improvement.* CSIRO Information Services, 314 Albert Street, East Melbourne, Victoria 3002, Australia.

Zar, J.H. (1974). *Biostatistical analysis,* 3rd Edn, Prentice-Hall International, USA.

Notes:

(1) See *Forest Genetic Resources,* published by FAO, Rome, Italy; and sheet D 70 in Manual 4 for more sources on genetic conservation of tropical trees;
(2) See sheet D 71 for more sources on agroforestry, mycorrhizas, nitrogen-fixing species, and organisations providing information;
(3) See sheet D 72 for more sources on formal and informal experiments.

Check-lists, sources and records - *estimating quantities* C 63

Introduction:

Unless your tree nursery is very small, setting it up and running it are likely to involve estimating:

(A) how many young trees will be needed (Manual 5);
(B) the nursery space needed for the different types of growing environments;
(C) how big the nursery should be; ***and***
(D) the amounts of various items that will be needed.

If you under-estimate, the work is likely to be held up, and insufficient numbers of young trees produced;
If you over-estimate, time and materials may be wasted, and too many plants produced.

This sheet gives some hints for making reasonably accurate predictions of what is likely to be needed.

(A) Estimating the number of trees to be grown:

(1) List the different kinds of trees and shrubs that are of interest.
(2) Estimate how many plants of each are likely to be needed, and when they are to be planted.
(3) Then increase the numbers to allow for losses, because:
  (a) the percentage of seeds germinating and cuttings rooting will be less than 100%;
  (b) some young trees may die during propagation, or be culled as unsuitable for planting; ***and perhaps***
  (c) it may be necessary to replace some planted trees that fail to establish.
(4) Work out the approximate totals to be grown in the nursery.

(B) Calculating the space needed:

(1) **For seed beds,** a rough estimate can be made by assuming that:
  (a) *larger* seeds should be sown at around *twice their diameter apart;* ***and***
  (b) *smaller* seeds should be spaced on average *no closer than 5 mm to each other.*

Then each square metre should produce approximately the following number of seedlings if they are well looked after (Manual 2); and the space needed for 1000 plants would be:

Seed diameter (mm)	Number of plants per m^2 of bed	Amount of space (in m^2) needed for 1000 plants assuming a germination percentage of 75 %	50 %	25 %
1	26,500	0.05	0.07	0.15
3	12,000	0.08	0.16	0.33
5	4,200	0.32	0.48	0.95
8	1,700	0.78	1.2	2.4
10	1,175	1.1	1.7	3.4
15	600	2.2	3.3	6.7

Seed trays have the advantage of being movable, but need around 25% more space.

When larger seeds are sown directly into a pot, the space may be calculated as in (B 3).

(2) **For cuttings** (Manual 1), assume that:

(a) *leafy cuttings* will need to be spaced in the polypropagator so that the leaves are not touching each other; ***and***

(b) *leafless cuttings* (if rooted in the nursery) should be placed in the propagation bed with spaces between them that are at least twice the diameter of the stems. Then:

Spacing of cuttings (cm x cm)	Number of cuttings per m²	Amount of space needed for 1000 cuttings (m²)
2 x 2	2500	0.40
3 x 3	1090	0.92
4 x 4	625	1.6
5 x 5	400	2.5
7.5 x 7.5	170	5.9
10 x 10	100	10.0
12.5 x 12.5	64	15.6
15 x 15	44	22.7

(3) **For standing ground for young trees in pots:**

Widest diameter of pot when filled (cm)	Number of pots across an 80 cm wide standing area	Running length of bed to hold 100 pots (m)
5	16	0.65
10	8	1.3
15	5	3.0
20	4	5.0

More space may be needed to allow enough room for trees with bushy shoots (C 42), or if the nursery has to be put on a steep slope (C 20).

(4) For **transplanting into beds** instead of using containers, the table in (B 2) can be used to estimate the space needed. Choose:

(a) *wider* spacings for trees that are to be planted as soil blocks (to leave enough room for the roots to be pruned), and for striplings and bare-rooted planting stock; ***but***

(b) *narrower* spacings for stumps (since both roots and shoots will be heavily pruned).

(C) Considering the total area needed for the tree nursery:

Add together the estimates for each of the growing areas. Multiply by a safety factor of 1.25. Then assume that you will need as much space again for paths, roads, buildings and so on, so double the figure to give a rough idea of the total size of the nursery.

For example, if you needed:

(1) 25 m² for seedbeds, 5 m² for seed trays, and 50 m² for large seeds sown into pots;

(2) 20 m² for polypropagators, and 15 m² for propagating leafless cuttings;

(3) 900 m² for standing ground for pots, and 335 m² for transplant beds;

then this gives a total of 1350 m². Twenty-five percent of that is 338 m², making a total of nearly 1700 m² for all the growing areas. The whole nursery might then occupy about 3375 m² (0.34 ha), not including room to expand. On the other hand, less space will be needed if different species can occupy the same growing areas at different times of year.

(D) Estimating how much potting mix will be needed:

(1) **In each container:** the volume of soil that different containers hold can be estimated in two ways:

(a) by closing the holes in the bottom of a pot of each size (C 6), and filling them up with water to the proper level for the top of the soil (C 42). Pour the water from each pot into a *measuring cylinder* to find out how many millilitres (ml) of water each contains. This is a rough estimate of the number of cubic centimetres (cm^3) of firmed down soil they will hold;

or

(b) by measuring the diameter and the height up to the soil level, and then calculating the volume:

For roughly cylindrical pots: divide the diameter by two to get the *radius.* The soil volume equals the radius squared, multiplied by 3.14, and multiplied by the height.
Volume = $\pi r^2 h$.

Example (a): if the diameter is 6 cm and the height 7 cm, then the soil volume is about 250 cm^3. *(Smaller than this, pots are only suitable for very small trees.)*

Example (b): if the diameter is 14 cm and the height 16.5 cm, the soil volume will be about 2500 cm^3. *(Larger than this, cylindrical pots are very heavy to use.)*

For tapered pots: divide the diameters at the top (soil level) and bottom by two to get the top radius and the bottom radius. Then the soil volume equals the top radius squared, plus the top radius multiplied by the bottom radius, plus the bottom radius squared; then multiplied by 3.14, then multiplied by the height and divided by 3.
Volume = $\pi^h/_3(r_1^2 + r_1 r_2 + r_2^2)$.

Example (c): if the two diameters are 13 cm and 10 cm, and the height 11 cm, then the soil volume is about 1150 cm^3.

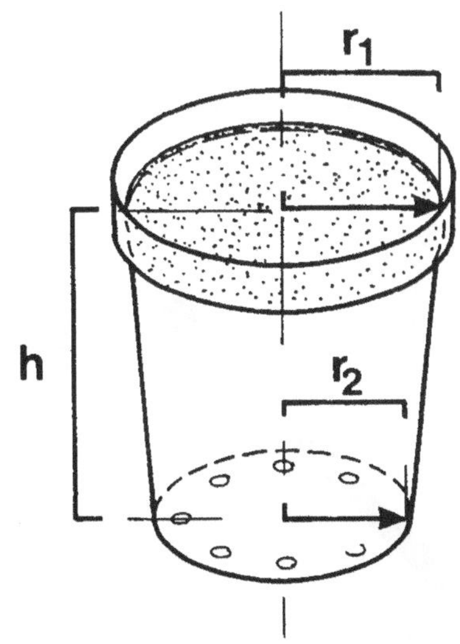

$$V = \pi \frac{h}{3}(r_1^2 + r_1 r_2 + r_2^2)$$

(2) **To pot up 100 young trees:** In (D 1) above, the volume of potting soil required will be about 0.025 m^3 for Example (a); 0.25 m^3 for Example (b); and 0.115 m^3 for Example (c).

(3) **To pot up 10,000 trees a year:** then the amount of the components needed for potting mix B1 in sheet C 6 for the 14 x 16.5 cm pots in example (b) above would be about 5 m^3 of coarse sand, 10 m^3 of loamy topsoil, 7.5 m^3 of weathered sawdust and 2.5 m^3 of compost each year.

(E) What other points need to be taken into account:

(1) You will also need reliable sources of good seeds (C 5 and Manual 2) and enough stock-plants to supply the shoots to make into cuttings (Manual 1).
(2) You might consider the potential for sale of unwanted planting stock.

See sheet C 61 for further information about nurseries.

1
2
3
4
5
6
7
8
9
10
1

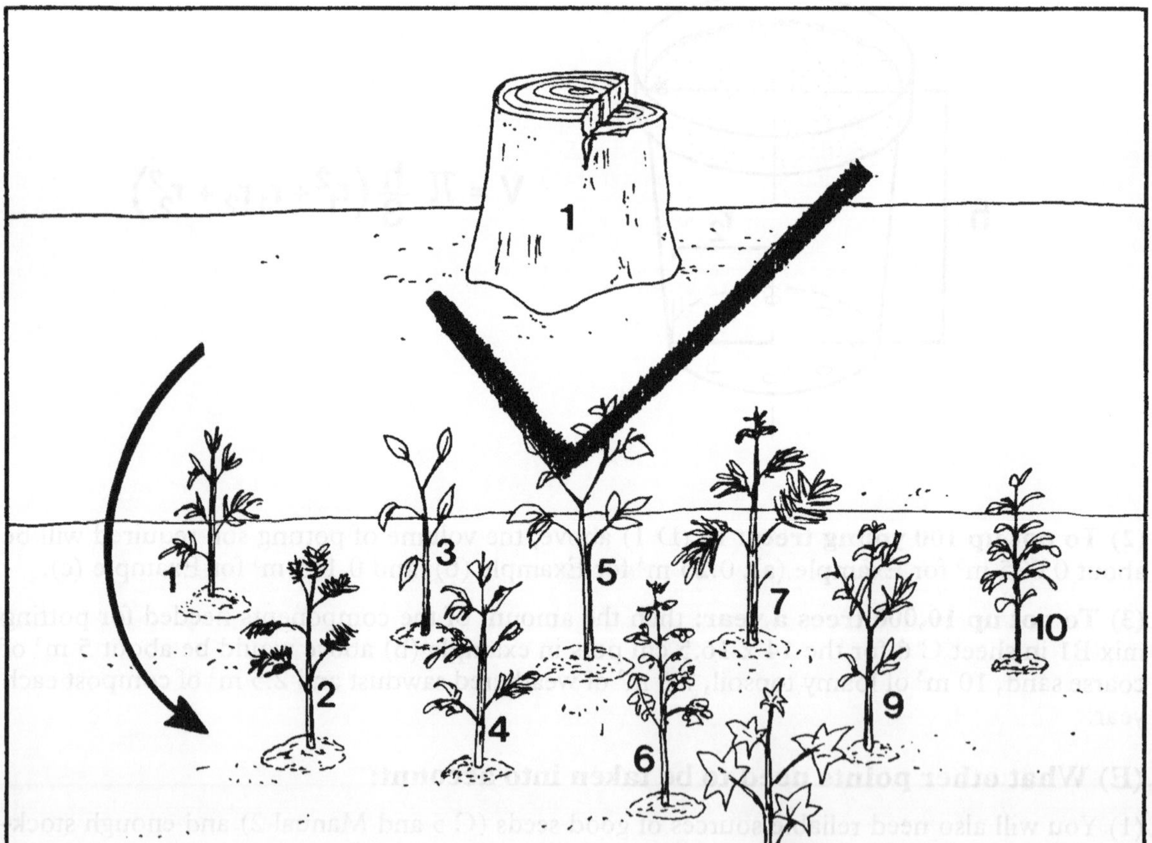
1
1
2
3
4
5
6
7
8
9
10

Check-lists, sources and records C 64

- record sheet for seeds, cuttings and plants collected or received

Identity number	Species	Type of material	Quantity	Date collected	Date received	Origin	Notes
Examples:							
98/1	Leucaena leucocephala	Seeds	250g	12/97	4/1/98	Ibadan, Nigeria	scarify (hard seeds)
98/2	Triplochiton scleroxylon	Cuttings	120	5/1/98	6/1/98	Mbalmayo, Cameroun	20 x 6 clones
98/3	Lovoa trichilioides	Plants	85	(1997 fruiting)	9/1/98	local forest	wildings

Check-lists, sources and records C 65
- record sheet for batches of plants grown

SPECIES: **IDENTITY NUMBER:**

PLACE OF ORIGIN OF SEEDS OR CUTTINGS:

Date of collection - / / Collected by -

Country - Provenance/Land race -

Exact locality - Approximate altitude -

SEEDS SOWN:

Date - / / Approximate amount sown - Where propagated -

Seed beds or seed trays? Germination medium -

Germination: very good/good/moderate/poor/nil, after weeks (% germinated)

Approximate number potted/transplanted - after weeks (% survived)

CUTTINGS SET:

Date - / / Approximate numbers of each clone -

Location of stockplants - Height at which cuttings taken -

Approximate length of cuttings - cm. +/- Auxin?

Where propagated - Rooting medium -

Number rooted - on / / (% rooted)

Number potted - on / / (% potted)

INTENDED USE OF THIS BATCH OF PLANTS:

For planting out - at site -

For potted plant experiments - For other research use -

For stockplants - To grow larger -

Other purposes -

Check-lists, sources and records C 66

- record sheet for checks made during nursery propagation

WEEK STARTING:

DAILY CHECKS:

	Monday	Tuesday	Wednesday	Thursday	Friday	Saturday	Sunday
Appearance of young trees:							
Wilting seen?							
Soil too wet?							
Leaves discoloured?							
Several leaves falling?							
Insect damage?							
Other animal damage?							
Growing conditions:							
Water supply in order?							
Shading intact?							
Weeding needed?							
Breaks in hedges or fences?							
Animal droppings?							

WEEKLY CHECKS:

Is each batch of young trees surviving and thriving?
Are any plants damaged, dead or missing?
Is it time to reduce any of the shading?
Are some batches ready for potting/transplanting?
Are weeds, insects and disease being kept in check?
What problems were found during the week?
Were they successfully dealt with?

Check-lists, sources and records C 67

- record sheets for measurements and analyses of variance

Note:

Page 210 is a blank assessment sheet for measuring trees.

Page 211 is the same sheet, with a simple worked example.

Page 212 is a blank analysis of variance sheet.

Page 213 is the same sheet, with a worked analysis of the figures on page 211.

ASSESSMENT OF ____________________

SPECIES :

EXPERIMENT NUMBER:

IDENTITY NUMBER:

DATE OF TREATMENT:

DATES OF ASSESSMENT:

DETAILS OF MEASUREMENTS MADE:

DATES →													
TREATMENT ↓	TREE NUMBER ↓												OBSERVATIONS AND NOTES
TOTAL													
NUMBER													
MEAN													

ASSESSMENT OF Gain in height growth

SPECIES: Ceiba pentandra

EXPERIMENT NUMBER: 3/97

DATE OF TREATMENT: 15/12/97

IDENTITY NUMBER: 96/12

DATES OF ASSESSMENT: 15/12/97 5/1/98

DETAILS OF MEASUREMENTS MADE:

Height in cm measured from the top of a peg in the soil to the estimated position of the main shoot tip

DATES →		15/12/97	5/1/98											OBSERVATIONS AND NOTES
TREATMENT ↓	TREE NUMBER ↓	Initial height	Height	Gain										
1 (small	1	20	24	4										
pots)	2	18	26	8										
	3	21	23	2										
	4	15	25	10										
	Total	74	98	24										
	Number	4	4	4										
	Mean			6.0										
2 (medium	5	19	25	6										
sized pots)	6	17	26	9										
	7	20	28	8										
	8	(22)	—	—										attacked by aphids
	Total	56	79	23										
	Number	3	3	3										
	Mean			7.7										
3 (large	9	18	32	14										
pots)	10	20	31	11										
	11	(17)	—	—										broken off
	12	(21)	—	—										died back
	Total	38	63	25										
	Number	2	2	2										
	Mean			12.5										
	TOTAL	168	240	72										
	NUMBER	9	9	9										
	MEAN			8.0										

Species: Experiment number:

ANALYSIS OF VARIANCE

Effect of:

Assessment of: Units:

Date of treatment: Date of assessment: Assessment number:

Treatment number → / Treatment →							OVERALL TOTALS
Total (Σx)							TOTAL (ΣX)
Number (n)							NUMBER (N)
Mean ($\bar{x}$)							MEAN ($\bar{X}$)
Differences: size: signicance:							C.F. = $\frac{(\Sigma X)^2}{N}$
$\Sigma(x^2)$							(A)
$\frac{(\Sigma x)^2}{n}$							(B)

Source of Variation	Sums of Squares	Df	Variance Estimate (Mean square)	Variance Ratio (sign.)	F (tables) (to exceed) (at level)
.					
.					
.					
.					
.					
.					
Treatment	(B − CF)				
Error	(total − treatment)				
Total	(A − CF)				

$LSD = t_{(\text{error Df})} \sqrt{\text{error variance} \times \left(\frac{1}{n_1} + \frac{1}{n_2}\right)}$

Coefficient of variation: %

Standard error of the mean:

at the
5%
1%
0.1%
levels

CONCLUSIONS:

Species: *Ceiba pentandra*

Experiment number: 3/97

ANALYSIS OF VARIANCE

Effect of: pot size

Assessment of: gain in height — Units: cm

Date of treatment: 15/12/97 — Date of assessment: 5/1/98 — Assessment number: 1

Treatment number →	1	2	3				OVERALL TOTALS	
Treatment →	small	medium	large					
Total (Σx)	24	23	25				TOTAL (ΣX)	72
Number (n)	4	3	2				NUMBER (N)	9
Mean ($\bar{x}$)	6.0	7.7	12.5				MEAN ($\bar{X}$)	8.0
Differences: size: signicance:	1.7 (1–2); 6.5* (1–3)	4.8 (2–3)		Treatment 3 = 2.1 x Treatment 1			C.F. = $\frac{(\Sigma X)^2}{N}$	576
$\Sigma(x^2)$	184	181	317				(A)	682
$\frac{(\Sigma x)^2}{n}$	144.00	176.33	312.50				(B)	632.83

Source of Variation	Sums of Squares		Df	Variance Estimate (Mean square)	Variance Ratio (sign.)	F (tables) (to exceed) (at level)
.						
.						
.						
.						
.						
.						
Treatment	(B – CF)	56.83	2	28.42	3.47 n.s.	5.14 (5%)
Error	(total – treatment)	49.17	6	8.194		
Total	(A – CF)	106.00	8			

$$LSD = t_{(error\ Df)}\sqrt{error\ variance \times \left(\frac{1}{n_1}+\frac{1}{n_2}\right)}$$

at the
5% — 6.1
1% — 9.2
0.1%
levels

Coefficient of variation: 35.8 %

Standard error of the mean:
for 4 replicates = ± 1.4
for 3 replicates = ± 1.7
for 2 replicates = ± 2.0

CONCLUSIONS: Overall treatment effect not significant. Growth in large pots more than twice that in small pots. This probably significant difference needs further study with many more trees in each treatment.

Check-lists, sources and records - *assessment by scoring* C 68

(A) Need for scoring methods:

Scoring is a valuable way of getting a rapid, general view of a situation in biology. It can take one further than the recording of an observation (C 55), without embarking on long and detailed measurements that may or may not be appropriate and productive.

Scoring is especially useful when the features to be assessed are difficult or impossible to record by measurement or counting; for example when differences are:

(1) primarily **qualitative**, such as the rooting of cuttings, the germination of seeds, or the stopping or re-starting of shoot elongation (C 12, C 55);
(2) rather **subjective,** like the branching habit of a young tree, or the colour of its leaves; ***or***
(3) needing to be estimated **without sacrificing** sample leaves (for example to measure leaf areas) or whole plants (to obtain dry weights).

Scoring can also be helpful later on, when the trees are **too big** for easy measurement, or there are **too many** items to count.

(B) How to score:

(1) Decide on the **feature** of the trees which you want to assess;
(2) Choose a set of recognisable **categories** or **stages** that cover at least the range of variation shown, and which can be seen without disturbing their growth;
(3) Label an example of each category or stage, and number them in sequence;
(4) Try giving a score to a few plants, and modify the categories if needed; ***and then***
(5) Score all the trees in the experiment (see the blank record sheet in C 67).

Suggestions on scoring leaf colour are given in sheet C 55.

(C) Some weaknesses of scoring methods:

(1) It can be difficult to standardise the categories, and the intervals between them are not necessarily equal;
(2) Bias (C 15) is harder to avoid than when measuring;
(3) After some time, one's brain may refuse to carry on scoring without a rest;
(4) Not all statistical tests (C 67, C 69-B2) can be done on the results, and extra care is needed not to mislead oneself.

(D) Hints on scoring:

The main aims when choosing a scoring method are to minimise these weaknesses, and to achieve a valid, useful assessment simply and promptly. Some hints are:

(1) Look through the young trees first in order to discover whether they have yet reached a suitable stage of development for scoring, and to gauge the range of variation to be expected in the feature(s) to be scored;
(2) Choose between 5 and 10 convenient categories or stages. For example, when assessing:
  (a) **categories** of *branching habits,* 1 might be used for young trees with unbranched main stems, and 5 for very bushy trees; ***and***
  (b) **stages** in *outgrowth of new shoots,* 1 might stand for "buds still unopened", and 10 for "first new leaves fully expanded";
(3) Do the scoring with at least one other person - difficult features may need three or four observers. Discuss the categories or stages together, but then score independently;
(4) Aiming for consistency of scoring is more important than whether you tend to score higher or lower than other people;
(5) Don't try and score too many different features at the same time; ***and***
(6) To reduce bias in experiments, arrange to do the scoring without knowing the treatment or genetic origin of the trees. Expectations can influence results!

(E) Analysis of scored data:

(1) Chi-square (χ^2) tests are especially appropriate for comparing categories. The 2×2 χ^2 test is quick to calculate, and gives a simple estimate of the *significance* (see C 69-E, G, H, I) of qualitative differences such as the presence or absence of something. For example: Comparing the number of cuttings that rooted in two different rooting media:

Treatment	Number rooted	Number not rooted	Total number	Percentage rooted
sand	19 (a)	46 (b)	65 (g)	29 %
sawdust/sand	12 (c)	3 (d)	15 (h)	80 %
both	31 (e)	49 (f)	80 (N)	

This is the sum for calculating the chi-square:

$$\chi^2 = (bc-ad)^2N/efgh$$

Note: when the numbers are small (some totals less than 30), three points apply:

(1) the test is only valid when the 'Expected Frequency' in each of the 4 main boxes a-d is at least 5 *(for example the expected frequency in box d = fh/N = 9.2 and so is valid);*

(2) 'Yates's Correction' should be applied before doing the test *(take ½ from higher value and add it to smaller); **and***

(3) the chi-square will only be significant if the difference between the percentages is quite large *(for example, a difference of 30 percentage points will not be significant until the total in each sample exceeds 25).*

Result of the chi-square analysis (using Yates's Correction): $\chi^2 = 11.18$ ***.

With one *degree of freedom* (see C 69-I) between the two treatments, the values of chi-square to be exceeded are 3.84 (5% level); 6.64 (1%); and 10.83 (0.1%). In this example, the difference in rooting percent is highly significant (C 69-H), indicated by ***.

When features have been scored into several categories, larger tables can be constructed, and the combined chi-square calculated. If such a test would be invalid because of low 'expected frequencies' in some boxes in the table, then categories can be amalgamated and a simpler table prepared. *(For instance, categories 1+2 and 3-5 might be put together and the larger groups compared in a 2×2 chi-square test.)*

(2) **Analysis of variance** can also be applied to scored data, and the variation between independent observers included in the analysis (see C 69-F), provided that:

a) the intervals between categories are reasonably even;

b) an appropriate *transformation* (see C 69-O) is used because the numbers are expressed in a small number of discrete categories, and may not be 'normally distributed' (C 69-B,2,g). Transformations may also be relevant when the categories are non-linear (for example, with numbers of branches in categories of 0, 1, 2-5, 6-14, 15+).

If there are many zero values, you could compare the presence or absence of the feature by a chi-square test, and confine the analysis of variance to the cases where the feature is present.

(F) Summary:

Used with judgement, scoring methods can provide a rapid and useful complement to more precise and fully quantitative measurements. They are especially valuable when time is short and the features do not lend themselves to easy measurement.

Although the data obtained are only semi-quantitative, it may be possible to carry out valid statistical tests of significance.

Check-lists, sources and records C 69
- *analysing the results of experiments*

(A) Why experimental results generally need analysing:

Looking carefully at what happened in your experiment can clarify:

(1) whether there were differences between any treatments or genetic origins;
(2) when they started to occur, and how big they became; ***and***
(3) possible linkages between observations and measurements, or between different assessments (C 55).

Statistical analyses are particularly important, helping one to **avoid being misled** when drawing conclusions about the results. They indicate how likely it is that any differences between the growth of various groups of young trees in your experiment are due just to chance, rather than to the conditions being studied (C 62-F).

(B) Two questions before starting a statistical analysis:

(1) Is it unnecessary? A formal analysis may not be needed for example when:

(a) numerous treated plants are thriving, while all the controls remain stunted;
(b) the trees of one genetic origin are growing, and those of the other are all dead; ***or***
(c) you were just doing a preliminary 'look-see' trial with a few plants, to try something out before a full experiment.

(2) Would the analysis be valid? It may not be, if, for instance:

(a) there were no controls or other standards to compare with the treated trees;
(b) the treatments were not applied randomly, or otherwise without bias;
(c) one treatment influenced another (for example if fertiliser in treated containers could have washed out and been taken up by control trees);
(d) a suitable layout of the young trees wasn't used during the experimental period;
(e) some of the experimental trees were subjected to severe stress (C 41) before the time of the assessment;
(f) the figures were arithmetically invalid *(for instance, ratios of percentages)*; ***or***
(g) the pattern of variation between the individual plants didn't approximate to a normal distribution *(but see section O)*.

(C) Steps in analysing the results:

(1) Calculating the **average values** for all treatments, Blocks *(D 55 in Manual 4)*, genetic origins, and other factors; from the measurements you have done, at each assessment;
(2) Finding out what **patterns of variation** occur - from selected samples of the individual values - and looking at how one set of figures might be linked to another;
(3) Preparing the main results as a **graph,** histogram, pie-chart, diagram or table, so that differences between sets of trees can be more easily seen and appreciated;
(4) Doing some **tests of significance** on the most relevant figures; ***and***
(5) Drawing **conclusions** about what the experiment has shown.

(D) Which figures to analyse?

This depends on the circumstances, but it may often be best to start with:

(1) assessments made at the end of the experimental period;
(2) stages when sizeable differences had recently appeared between various sets of experimental trees; ***or***
(3) differences that deal with the main hypotheses of the experiment.

A decision can then be taken about which other sets of figures might be worth analysing.

(E) Tests of significance for 'Yes/No' situations:

If the difference between two sets of experimental trees is **qualitative** - that is, a simple choice between damaged/undamaged; alive/dead; leafy/leafless; terminal bud sprouting/not sprouting - then the **Chi-square test** (χ^2) is a straightforward one to use.
(see worked example of a 2×2 χ^2 test in sheet C 68-E.)
Chi-square tests can also be performed with more than two samples.

(F) Tests of significance for 'More/Less' situations:

Where the difference between various sets of experimental trees is **quantitative**, several kinds of tests of significance are available. One of the most adaptable and widely used is the **Analysis of Variance** (ANOVA) - see blank sheet and worked example in C 67). What this does is to estimate:

(1) how much variation exists between all the individual values being analysed;
(2) how much of this variation can be assigned to the treatments applied, to overall differences between Blocks, to various genetic origins, or to other factors; ***and***
(3) how much then remains unassigned (the 'Error' or 'Residual' estimate).

A simpler version is the **t-test,** but this can only handle a single comparison at a time, so is generally less informative and useful. *(See also J - standard error of the mean.)*

(G) Significant and non-significant effects.

The starting assumption *('null hypothesis')* on which tests of significance are based is that there are <u>not</u> any real differences between the various groups of plants in the experiment - they just show chance variation around the overall mean. However, if it turns out that considerably more of the variation is assigned to treatment than to error, the assumption is found to be false and the treatment is said to have had a *significant* effect. If, on the other hand, roughly similar variation is assigned to treatment and to error, the original assumption stands, and the overall treatment difference is said to be *'not significant'* (n.s.). The same applies to Blocks, genetic origins, and other factors.

(H) Levels of significance.

If a test of significance gives a number that is **larger** than the value given in the relevant table for the 5% level of probability ($p = 0.05$), this means that variation like this might happen anyway by chance in one out of more than 20 such trials. We say that such a difference is *"probably significant"* (and it is usually given one *). If the test gives a number that is bigger than the value in the table for the 1% level ($p = 0.01$), such a difference would only be likely to happen by chance once in more than 100 trials, and it is called *"significant"* (**). If the value for the 0.1% level ($p = 0.001$) is exceeded, the difference would probably only occur by chance once in more than 1000 trials, and is called *"highly significant"* (***).

(I) Degrees of freedom and significance in statistical tables.

Between two treatments, there is only one comparison to be made; between ten seed-lots only nine independent comparisons. The number of degrees of freedom (d.f.) is the number of trees, replicates, Blocks, treatments, seed lots, clones, and so on that are involved; minus one. So when looking up tables for:

(1) *Chi-square tests:* With one d.f. between 'yes' or 'no', the values of chi-square to be exceeded are 3.84 (5% level); 6.64 (1%); and 10.83 (0.1%).

(2) *ANOVA* (See worked example in C 67):

(a) Divide each of the *sums of squares* by the appropriate d.f. to get the *mean squares* (estimates of variation);
(b) Divide the mean square for treatment by the error mean square, to get the calculated value of F *(variance ratio);*
(c) Look up a Table of F values, using the d.f. for treatment along the top, and the error d.f. down the side. If your calculated F is larger than the one in the Table, this means that there is a significant **overall effect** of treatment *(see J and L for individual pairs of treatments);*
(d) Similarly, the mean squares for Blocks, clones, 'triplets' (C 15) or other groupings are divided by the error mean square to find out whether any of them show significance.

(J) Calculating the Standard Errors of the Means:

The standard error of the mean (S.E.) is the simplest estimate of how reliable an average is. It can be calculated for any set of figures by dividing the standard deviation by the square root of the number of trees:

$$\text{S.E.} = \frac{\text{standard deviation}}{\sqrt{n}}$$

After an ANOVA, a more accurate S.E. is calculated by dividing the error mean square (residual variance estimate - see I-2) by the number of values that have been averaged in a particular treatment, and then taking the square root.

$$(\text{S.E.} = \sqrt{\frac{\text{error mean square}}{n}})$$

The average (mean) is then written for example as 5.6±1.2, and on a graph or histogram the S.E. is usually shown to scale, as a vertical bar above and below the average value.

If their 'error bars' do not overlap, this is commonly taken as an indication that two means are probably significantly different from each other. If they do overlap, any differences may just be due to chance.

(K) Interactions.

Consider a 2 x 2 trial with a control, mulch only, fertiliser only, and mulch plus fertiliser (D 6 and D 55 in Manual 4). If, for example the effects of fertiliser were dependent on whether the plants were mulched or not, then an *interaction* is occurring (Manual 5). The two factors, mulch and fertiliser, are not acting independently from each other. Interactions are important in understanding more about growth, because they suggest that the two factors are acting upon the same process. On the other hand, if there is no interaction, the separate effects of the two factors will just be added together, or the one subtracted from the other. Interactions:

(1) can only be detected in experiments examining more than one factor. These might for instance be two different types of treatment, or one kind of treatment and a difference of genetic origin;
(2) need to be considered before looking at the effects of the main factors on their own;
(3) may, if significant, mean that a 'breakdown analysis' (usually, re-analysing the experiment in two parts) is needed to determine whether the individual effects are also significant.

(L) Examining the significance of differences between individual pairs:

(1) Find **the difference** between the average values in the pair to be compared *(for example between treatments 1 and 3 in the worked example in C 67);*
(2) Calculate a value called the ***least significant difference*** (L.S.D.):

L.S.D = t x the standard error of the difference.

$$\text{L.S.D.} = t\sqrt{\text{error mean square} \times \left(\frac{1}{n_1} + \frac{1}{n_2}\right)}$$

The value 't' is taken from tables at the 5%, 1% and 0.1% levels of probability, using the **error** degrees of freedom. n_1 and n_2 are the numbers of plants in treatment 1 and control;
(3) If the difference between the two averages is larger than the L.S.D., the difference is significant at the appropriate level.

This is the simplest method. Although various authors suggest alternatives (C 62-F), these are more complicated to calculate. The L.S.D. can be a useful guide, provided the following points are remembered:

(a) if 20 genetic origins were tested in an experiment, you would expect the LSD at the 5% level to be exceeded once, just by chance, not because there was a real difference;
(b) similarly, if the treatments in an experiment involved 3 factors at 3 different levels, at least one of the 26 possible comparisons would be expected to be significant at the 5% level, without meaning that a real difference had been found.

(M) Various reasons for lack of significance:

If a test does not show significance, this merely means that the null hypothesis stands (see G). It **does not** prove that the treatment is ineffective.

Significant effects might not have been found because:
(1) there was too much variation in the experiment for them to show up;
(2) there weren't enough replicates, as in the worked example in C 67;
(3) the experimental plants were growing slowly, and errors in measurement were too large for an effect of treatment to be found;
(4) the effect of another factor was masking that of the treatment in question; ***or***
(5) the particular treatment would only show significant effects in a different environment.

(N) Reducing variability:

(1) *Re-analysing data from the same experiment:*
(a) You could recalculate the data as the gain since the start of the experiment (C 55). This removes the variation due to the trees starting off at different sizes, and is in any case desirable when studying the periodicity or rates at which shoots are growing;
(b) For *relative* growth rates, you could do an analsis of *covariance,* based on the values for individual trees at the beginning of the experiment, and after a given period, or a regression (see Q);
(c) Transformations (see O) may have the effect of reducing variation, because they often give less weight to occasional very high values;
(d) Some computer programmes (see R-2) can be set to ignore data points that are further from the mean than a set distance. This is risky when dealing with variable species and environments, as these points may well be true values.

(2) *Repeating the experiment.* For instance, you might do this using:
(a) more replicates;
(b) young trees that had been grown beforehand under more uniform conditions (C 7);
(c) a different experimental area that provided similar light levels to all the plants. If necessary, you could have 3-5 Blocks running from the sunnier to the shadier parts;
(d) a 'surround' of similar trees that were not part of the experiment, to reduce 'edge' effects;
(e) treatments that were more contrasting than before; ***and***
(f) more careful handling and watering (C 42, C 48)

Only after several experiments would you conclude that the treatment probably has little or no effect on those aspects of growth of that tree species.

(O) Transformations.

These are sometimes needed in order to put the figures into a form where a valid analysis can be done (see B-2). Here are some examples:
(1) Chi-square tests with small numbers - apply Yates's Correction (see C 68-E).

(2) ANOVA with a non-normal distribution - if the mean value lies well towards the low end of the distribution, transforming all the original data ('x') may make the distribution reasonably normal. If so, do the ANOVA on the transformed figures ('z'). Some common transformations include:
(a) square root transformation: $z = \sqrt{(x + 0.375)}$
(b) log transformation: $z = \log^{10} (x + 0.375)$; or $z = \ln (x + 0.375)$
(0.375 is added to each number to avoid problems with values of zero and one.)

(3) ANOVA of percentages - use the arcsin transformation:

$$z = \sin^{-1}\sqrt{\frac{x\%}{100}}$$

Note: if you want to de-transform the results before presenting them, the standard error of the mean (see J) and the least significant difference (see L) require care. Because transformations (2) and (3) above are not linear ones, the S.E. bars will be of unequal length above and below a mean.

(P) Missing plants or readings.

It is still possible to do ANOVAs when there are different numbers of readings in the various treatments or genetic origins, for example because:

(1) there was a shortage in some groups of plants;
(2) only a few trees could be treated, but more controls were available;
(3) some trees were accidentally damaged during the experiment;
(4) some dieback of shoot tips occurred, or death of plants (C 55).

If the ANOVA has only one factor (see K), then calculate as in C 67.
If it has more than one factor, you could analyse them separately, though without being able to look at any interactions. Alternatively see statistical textbooks (C 62-F) for how to estimate missing values, noting that for each of them one d.f is deducted before calculating the error mean square.

(Q) Correlation and regression.

These are ways of examining how closely two sets of readings may be connected. For example, height and diameter growth in a set of young trees might often (though not always) be closely linked, with the shorter trees thinner, and the taller ones thicker.
Moreover, you might expect that the growth of the trees could be linked with soil depth, moisture or fertility, or with an aspect of the weather.

When correlations or regressions show a close relationship, significance values are often given to them. But here it is particularly important not to be misled, because:

(1) the significance has not come from the experimental testing of a hypothesis;
(2) the apparent link may simply depend on connections to a third factor; ***and***
(3) one in twenty comparisons amongst the many factors affecting the young trees may be expected to show a 'probably significant' relationship just by chance.

(R) Aids to calculation.

(1) **Calculators:** These have the advantages of being small, portable, robust and relatively cheap, and of working reliably from long-lasting batteries or solar energy. They are invaluable for transformations (see O), to obtain and check totals and averages, and for other simple calculations (C 63).

Some types contain programmes that automatically calculate the standard deviation and standard error of the mean when a set of figures is totalled. Others will perform more detailed analyses, or allow you to write a programme yourself.

(2) **Computers:** These offer opportunities for storing large amounts of data, doing complex calculations and analyses, and almost limitless possibilities for displaying the results. They can also be programmed to accept electronically recorded information about the environment. However, computers are relatively expensive, require a steady and reliable electricity supply, and need to be kept free of dust and high humidity. Some types can operate from rechargeable batteries, but they are too delicate to be really portable.

(S) A final hint:

Check ***at each stage*** for errors in recording numbers, and in calculations. If not, sooner or later you will find yourself having to start back at the beginning, re-analysing and re-drawing graphs (and perhaps even changing slides and the proofs of publications). Just because a computer has done the analysis does not mean that there cannot be errors.

INDEX

Index

Pages containing guide-lines are shown in **bold**

Postscript

During the latter part of 1997, smoke, ash and gases from the burning of large numbers of tropical trees and peatlands led to widespread 'smog' in South-east Asia, which caused serious problems for the inhabitants.

Tree planting is now even more important than previously estimated, because:

- *(1) many fires were burning out of control;*
- *(2) few of the trees there will re-sprout;*
- *(3) few sources of seeds will remain in the burnt areas; and*
- *(4) many sites will be liable to become degraded unless they are weeded and planted with trees.*

Tropical Trees: Propagation and Planting Manuals
VOLUMES PLANNED FOR THE COMPLETE SERIES

Volume 1 - Rooting Cuttings of Tropical Trees

introduction
genetic selection
stockplant management
propagating conditions
taking the cuttings
care of cuttings
check-lists, sources and records

Published 1993
Accompanying video published 1994 by ECTF Penicuik EH26 0PH, Scotland

Volume 2 - Raising Seedlings of Tropical Trees

introduction
sexual reproduction in trees
choosing seed sources
seed handling
germinating the seeds
care of young seedlings
check-lists, sources and records

Expected 1999

Volume 3 - Growing Good Tropical Trees for Planting

introduction
general principles of tree growth
planning a tree nursery
micro-organisms, nutrition and tree growth
protecting growing trees
running a tree nursery
check-lists, sources and records

Published 1998

Volume 4 - Preparing to Plant Tropical Trees

introduction
general principles of tree survival
types of planting site
which tree species, for what purpose?
deciding on the growing system
preparing the ground
sources of further information

Published 1995

Volume 5 - Planting and Establishment of Tropical Trees

introduction
taking the young trees to the planting site
how and when to plant
protection
successful establishment
assessing the results of field trials
check-lists, sources and records

Expected 2001

Printed and bound by CPI Group (UK) Ltd, Croydon, CR0 4YY
16/09/2026
14963440-0001